Skulls

Skulls

An Exploration of Alan Dudley's Curious Collection

Simon Winchester

Skull Collection by Alan Dudley

Photography by Nick Mann

BLACK DOG
& LEVENTHAL
PUBLISHERS
NEW YORK

Published by
Black Dog & Leventhal Publishers, Inc.
151 West 19th Street
New York, NY 10011

Distributed by
Workman Publishing Company
225 Varick Street
New York, NY 10014

Manufactured in China

Cover and interior design by Matthew Riley Cokeley

ISBN-13: 978-1-57912-912-5

h g f e d c b a

Library of Congress Cataloging-in-Publication Data available on file.

Contents

Contents

Introduction

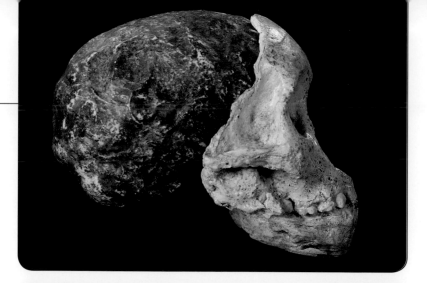

THE WONDERFULLY COMPLICATED mass of gray and pink nervous tissue that constitutes an animal's brain is a thing both entirely vital to the creature's existence and, in and of itself, an organ utterly frail and terribly fragile. It both needs and deserves a strong, hard case to surround and enclose it, to give it protection, and to serve as its housing and its frame.

That hard case, which evolutionary development has caused in most creatures to be made strongly—of bone—also holds and supports, in places that are conveniently close to the brain inside, the sense organs—those smaller and complex devices that allow the animal to see, to hear, to taste, and to smell. The case is also typically pierced by holes through which the animal ingests food and exchanges air or water with its surroundings, extracting oxygen and useful chemical signals, like those encapsulated in smells.

Usually the brain is placed at the leading end of the animal—at its front if the animal moves horizontally, at its top if it moves in an upright fashion. This has a singular effect: that the case which holds the brain, and on which are suspended the sense organs—the eyes, the ears, the nose, and the mouth, most often arranged into what is called a face—is usually the first part of the animal to be seen. One might say that, because of this, the leading end of the animal defines it, gives it its character, its uniqueness, its aspect.

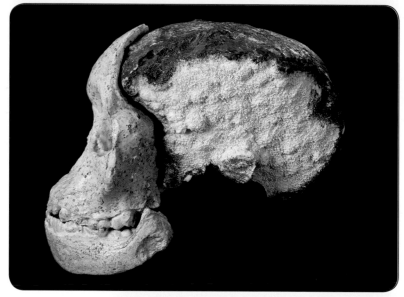

When it is covered with skin and muscles, with fat, blood vessels, with fur or hair, we recognize the entire confection as the animal's head. But when this head is stripped of all its coverings and its organs, when it's reduced to its basic foundations, we know what remains—one of the most familiar icons of the animal kingdom—by one of the most ancient words in the English language: we know it as the skull.

This book tells the story of the skull, in both the human and the animal world. Skulls—human in the main, but by no means exclusively—have exerted for scores of thousands of years an almost inexplicable power over the human imagination. Skulls are symbols both of existence and of former existence; they are freighted with terror and awe; they tell of life, death, and the afterlife; of good and evil; of danger, authority, and majesty. Perhaps no other biological entity retains such a grip on human psychology as does this assemblage of hollow bone, this thing of domes and sockets and jaws and of mysterious interior passageways and canals. People are fascinated and captivated by skulls—of animals and humans alike. We always have been and always will be.

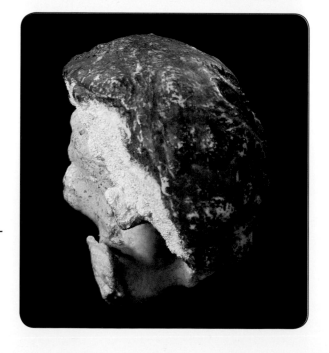

The Taung Child, pictured here from four different angles, is an infant example of *Australopithecus africanus*. Apart from being a wonderful find in its own right, this skull also contained an extremely rare natural endocast (a cast of the internal surface of the skull that occurred during the fossilization process). It demonstrates how the modern dentition and bipedal stance of the hominids evolved before the enlarging of the brain, which remains the size of a chimpanzee in this species. Indeed bipedalism is considered a prerequisite of brain enlargement as beyond a certain size the head becomes too heavy to support without the assistance of the spine.

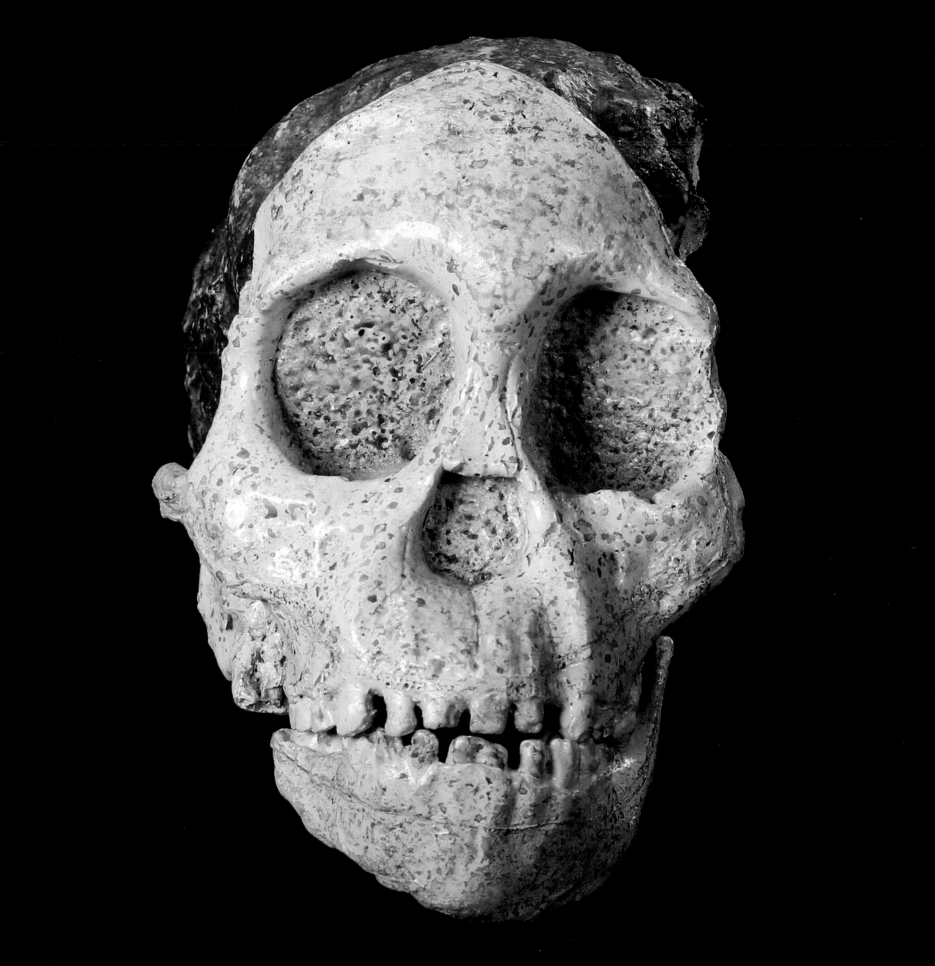

The Collector

ALAN DUDLEY is by training and profession one of a rare breed of specialists who select and treat wood veneers for use in the interiors of very expensive British motor cars. That's his day job. After work, he collects skulls.

As with many obsessive fascinations, his hobby began by chance. In common with most small English boys of his generation—Dudley was born in 1957, in the Midlands city of Coventry—he was fascinated with wildlife: he collected birds' eggs and kept newts in a jar. His interest was both stimulated and maintained by television, specifically by the endlessly infective enthusiasm of the near-legendary presenter of TV wildlife documentaries, David Attenborough.

And so, while the young Dudley was by no means obsessed with animals and birds—he would describe his interest as above average, but not all-consuming—it made perfect sense that when, one day soon after his eighteenth birthday, he found the carcass of a dead fox draped on a garden fence, he should take it home, with a view to cleaning and studying it.

By the time he collected it, the three-foot-long animal was little more than a ragged mess of fur draped around a core of skeletal bones—the flesh had all rotted away. And while seeing it in this condition gave him an initial insight into the possibilities of taxidermy (which he went on to do, using other animals, and became more than a little skillful and successful), with this particular specimen he dismounted the head, took off the fur covering with a knife and tweezers, and got his first good look at a whole skull, magnificent in its purity and perfection.

He learned the various ways of preparing a skull. Some collectors, once having performed the rough cleaning with knives, employ small armies of flesh-eating maggots, larvae of dermestid beetles, to nibble their way around the crevices and eat out the tiny remaining morsels of flesh.

But Alan Dudley found—in both his first fox and in a large number of other of the more delicately-boned skulls—that the maggots were too brutal in their treatment of some of the finer bones, pushing them aside (and thus distorting them) in their eating frenzy. Sheaves of tiny flat bones, for instance, within the deeper recesses of the noses of some long-nosed animals and birds, are often ruined. So Dudley's preferred method is cold water maceration: basically, soaking the new-found skulls in a bucket of water for a very long time and letting any still-adhering flesh dissolve away, with help from the bacteria that find their way into the bucket.

The process is extremely smelly (hence Dudley's preference to work in the garden shed) and very time-consuming (hot water works faster, but can damage the skull, causing the teeth to fall out and the brain to expand and split the braincase). But after some weeks or months submerged in water that becomes dark and perfectly horrible to be near, a skull is rendered clean and flesh-free. The blood vessels, the bands of cartilage and clumps of muscle, as well as the eyes and tongue and soft palate and hearing mechanisms, all vanish, and what remains is an off-white amassment of curvilinear bones, some hard and some soft, some massive and some delicate, that can be washed, whitened (with hydrogen peroxide—never, ever with bleach) and perhaps lightly varnished, to be labeled, identified, and placed in a display case for the remainder of time.

Over the years—as his single fox became a fox and a bat; then a fox and a bat and a newt; then fox, bat, newt, anteater, owl, cuckoo, monkey; and on and on and on—Alan Dudley became an extremely accomplished skull collector. He was well known in some circles, seen as an authority, revered as someone possessed of what was fast becoming an excellent reference collection of skulls, an amassment of great breadth and depth that had been prepared and labeled as if of almost museum quality.

He soon began dealing with nearby zoos—curators would call him if one of their animals died, so that he could, if he wanted, collect the head and bring it home for maceration, preparation, and placement. He began trading skulls with other collectors, both in Britain and with large dealers in the United States.

The Arrest

WHEN ALAN DUDLEY started trading skulls over the Internet, he was well aware of all the various international treaties and laws covering the exploitation of some kinds of animals. He had long supposed that his

> **Dudley found that the maggots were too brutal in their treatment of some of the finer bones, pushing them aside (and thus distorting them) in their eating frenzy.**

Alan Dudley's collection room.

collection conformed in all ways with legal requirements and restrictions, and he was well aware of the dangers and evils of poaching, aware that endangered species were regularly taken illegally and plundered for various parts—mainly hides, tusks, glands, and reproductive organs—that fetched stellar prices in particular corners of the world.

All of this he knew. Which is why Dudley was more than a little surprised when, one March afternoon in 2008, four police officers arrived at his door. One was a wildlife crime officer, one an enforcement officer from Customs and Excise, and the two others were uniformed constables (as escorts). They came with a search warrant, so Dudley had no choice but to let them carry out a lengthy inspection of his enormous collection.

While the police found that the vast majority of the collection was entirely legal, some specimens had clearly been bought in disregard of the law. Dudley was eventually charged with a crime: with seven counts of having breached CITES (Convention on International Trade in Endangered Species of Wild Fauna and Flora). The skulls that most exercised the courts were those of a howler monkey from Ecuador, a penguin, a loggerhead turtle, a chimpanzee, a Goeldi's marmoset from Bolivia, and a tiger. He was ordered to wear an electronic ankle monitor—sufficiently carefully calibrated that it prevented him from having access to his collection, which had crime-scene tape strung across the door. And when eventually the case went to Coventry Crown Court, where he pleaded guilty, he was sentenced to fifty weeks prison time, suspended, and the specimens that lay at the center of the controversy were confiscated.

The judge in the case—who spoke of an "academic zeal" that had "crossed the line" into "unlawful obsession"—remarked in passing sentence that the most egregious example of lawbreaking was Alan Dudley's purchase of the howler monkey skull. Pictures of three monkey skulls for sale had been posted on the Internet, and it was clearly apparent that two of the three had been shot in the head. That Mr. Dudley had bought the one specimen without an apparent head wound was no excuse, said the judge: "You must have known you would never, and I mean never, have received any license to import that skull. It must have been

absolutely, abundantly, crystal clear to you that the provenance of that trio of skulls was extremely dubious to say the least. No person involved in lawful activity would, in my view, purchase skulls from a protected species which showed that they had been shot through the skull. And two of the trio showed exactly that."

The judge imposed the suspended prison sentence, and fined Dudley £1,000, with £1,500 costs, for the offense.

Once reunited with his remaining collection—the ankle monitor removed, the crime-scene tape stripped away—Dudley vowed to be more cautious in his purchasing practices, less impetuous in deciding to buy some of the rarer species that he still needs to expand his world-class collection. On the day we met, he had been to the local fish market to acquire—perfectly legally—an angler fish, with its spectacular illuminated lures and a magically ugly face that overlays a fascinatingly shaped skull.

Dudley remains sturdily enthusiastic about his curious obsession. He reminisced about his earliest—and to outsiders, somewhat grotesque—triumphs. He tells of finding a dead fox cub curled up beside a ditch, and then spending hours wading in the cold and muddy water searching for a complete set of the animal's teeth (as with most dogs, a fox has forty-two); of finding the body of a tethered and starved Great Dane in a dilapidated apartment complex in Spain, and cutting off its head with a penknife; of his delight at finding a dead barn owl in a field; of the tortoise his mother had buried in a plastic bag and which had turned entirely to mush; of a hedgehog he found, with which children had been playing football and "yet with its nasal bones still not damaged by the roughness of their game"; of how his then-wife Jacqueline vowed to destroy his entire collection because of the dreadful smell of a rotting green iguana; and of how, when he saw a spider monkey that had died of cancer, "I couldn't bring myself to take her head, so sad was her story."

The Nature of Collecting

BY ALAN DUDLEY'S own admission, the collecting of skulls is a curious calling—though perhaps no more so than of many of other objects that excite a zeal for acquisition—a zeal that has long intrigued psychologists (and

novelists: John Fowles wrote *The Collector*, in which a butterfly collector adds an innocent woman to his haul, with predictably tragic results).

The difference between unorganized acquiring and hoarding on the one hand, and systematized acquisition and classification on the other, is often a pathological need to win psychological security through the domination of inanimate entities. But in most cases—stamp collecting, coin collecting, matchbox- and antique- and beermat- and vintage car-collecting—collecting is entirely harmless. Indeed, our propensity to collect is the basis of an immense industry. Skull collecting, though it may at first seem macabre and bizarre and even slightly frightening, is probably no different and has the added benefit of being entirely educational.

But it can lead, somewhat more easily than can other kinds of collecting, into somewhat unseemly behaviors and unsavory company. The pursuit of skulls has led to grave robbing, for instance; to poaching; to the hunting of animals that should be well protected. And it can also lead into intellectual byways that can prove to be highly dubious.

It led, for instance, the famous nineteenth-century American physician and scientist Samuel George Morton—who built up a collection of more than a 1,000 human skulls—to come up with an elaborate and now fully discredited theory that has helped to bolster much racist thought in the United States.

Morton's fascination with craniometry led him to measure his skulls in detail to determine their relative brain capacities. He believed that his findings supported the idea that humankind had always been comprised of five separate species, with Caucasians at the summit of human achievement and ability, the darker races, as he put it, "from the beginning" destined to serve as their slaves and servants. Southern whites flocked to Morton's side, believing he had proved a justification for slavery, and for many years he was the darling of all racist groups in the country. After his death—and he was a quiet, modest man, who shied away from the political implications of his work—his skull measurements were recalculated and shown to be correct. The theory he derived from them, however, never won widespread acceptance.

Dudley is a talented taxidermist and several examples of his work can be seen in his collection room. This is a red-tailed hawk, sometimes known as a chicken hawk. It is native to North America.

Notes on the Collection

THERE ARE MORE than 2,000 skulls in Alan Dudley's collection. Although it's tucked into a spare bedroom at the top of the stairs in a small house in the English Midlands, it's actually one of the largest and most comprehensive in private hands anywhere in the world—so rich, in fact, that it was really quite a challenge to make a selection for this book.

Our intention in doing so, in offering up a precise distillation of this remarkable assemblage, can perhaps best be summarized as think taxonomy. For while we were happy to include some of Dudley's personal favorites (he always said that if his house caught on fire, he'd first rush to save his rhinoceros hornbill skull, his hippo, his orangutan, and his southern sea lion and then go back into the inferno to scoop up the babirusa , the shoebill, his mandrill, the gorilla, the one duck-billed platypus, and, most bizarrely, his two-headed cow), we knew also that to make this compilation truly useful, we had to present as representative a cross section of the vertebrate universe as was possible, deriving our selection as much as we could from the skulls in Dudley's collection but supplementing it if necessary.

Our singular achievement—if, as we hope, this is what it turns out to be—was helped to a large degree by the quality and depth of Alan Dudley's collection.

It turned out to be easy enough, for example, to make selections that represent the world's 5,000 extant mammals: Dudley has fine examples from all three subclasses—the egg layers, the marsupials, and the placental mammals—and from most of the twenty-five-odd orders of the third and largest of these (the exceptions being only elephants and manatees).

There are approximately thirty-two orders of birds. Dudley has skulls from twenty or so of them. So while this meant that our selection would necessarily be a little circumscribed, we can fairly say that all the major avian types are represented.

The same is true—though for rather more complex biological and evolutionary reasons—for the three major taxonomic groupings of reptiles, all of which are well represented both in his source collection and here—plenty of snakes, lizards, crocodiles, and turtles.

There are only three orders of amphibians. Thanks to Dudley, we have two of these. The third, an obscure group of wormlike burrowing creatures called caecilians, are not represented.

We have selected as well as we can from the immense complexity of the fish world—a world of more than 30,000 species, many of them with skulls—or front ends, since often a fish has no real skull as such—of a daunting delicacy.

All told, then, since the taxonomic quality of the main collection itself is outstanding, so are the quality and scope of the selection from it that we offer here.

Moreover, we can fairly say that all of our chosen specimens have been as carefully and properly identified as possible—a task that was made reasonably easy as a result of knowing their provenance: the vast majority of Dudley's specimens came from zoos, whose staff clearly knew very well, and in great detail, what they had possessed.

The antique skulls of some of the rarer creatures, which Dudley has thus been obliged to obtain from dealers and other specialists, are also classified and named. Inevitably, there will be some courteous disagreement and argument—there always is among biologists when they consider the finer branches of the Tree of Life; but our firm belief is that each specimen presented here is as accurately identified, and the sample is in total as taxonomically comprehensive, as any known, anywhere.

There were a few gaps that Dudley couldn't fill. We've provided some images of skulls not found in his bedroom collection—some because they were too rare, others because they were too large, but all ones that Dudley would dearly love to own. Among these are the skulls of an elephant and its tusks, a rhino and its horns, and the famous Oxford dodo.

There are some curiosities from his collection that are not skulls—some tortoise shells and a magnificent-looking (but only a cleverly made and highly accurate simulation of) model skull of a saber-toothed tiger .

Finally, a note on the design. We have taken great care to present the skulls from their best and most interesting angles, while making every effort to display them in an organized manner based on class, family, and species. Some we have chosen to present as large as possible and others at a smaller size. They are not meant to be to scale.

What we have here in *Skulls* is, in other words, a near-perfect survey—an assemblage of biology that is balanced, spread wide, and taxonomically comprehensive—designed for browsing, for amusement, for macabre fascination, but, most importantly of all, for learning. We like to think that this set of images does full justice to the most emblematic component of this planet's vast menagerie of animals with backbones: a fair and nuanced account of the skull, the beautifully formed casing structure in which resides all the complex mechanisms and sense organs that make the world's higher animals truly the marvels that they (and we) are.

Hippo skulls are truly huge. Only elephants and white rhinos have skulls that are similarly massive.

Amphibians

North American Bullfrog
Rana catesbeiana

THE NORTH AMERICAN BULLFROG'S skull is fragile, with large holes for the eyes (which are retracted somewhat during walking and swallowing), and an immensely out-of-proportion mouth. The mandible is toothless, but the upper jaw has rows of tiny teeth.

Kingdom: Animalia
Phylum: Chordata
Class: Amphibia
Order: Anura
Family: Ranidae
Genus: Rana
Behavior: Insectivore/Nocturnal

Argentine Horned Frog
Ceratophrys ornate ▽

THE JAWS ON THIS AMPHIBIAN are so large it could just about swallow itself. This horned frog from Argentina—the skeleton displayed here as if about to strike—is a voracious eater that will seize on almost anything that passes by. Otherwise known as the wide-mouthed frog or pacman frog (after the famous video game), it has an impressively large skull for its size. Argentine horned frogs are often kept as pets and feed on mice.

AKA: Argentine
Wide-mouthed Frog,
Pacman Frog
Kingdom: Animalia
Phylum: Chordata
Class: Amphibia

Order: Anura
Family: Leptodactylidae
Genus: Ceratophrys
Behavior: Carnivore/
Nocturnal

Axolotl
Ambystoma mexicanum

MUCH LOVED by zoologists (and cunning Scrabble® players), the axolotl is a newt that never grew up. It sounds bizarre, but it is a giant newt that retains its juvenile gills. Zoologists call this neoteny: sexual maturity is reached before the animal is fully developed, effectively retaining juvenile characteristics into adulthood. This axolotl skull is very delicate, but typically amphibian, with hundreds of tiny pointed teeth and a huge mouth hinged at the rear.

Kingdom: Animalia
Phylum: Chordata
Class: Amphibia
Order: Caudata
Family: Ambystomatidae
Genus: Ambystoma
Behavior: Carnivore/
Nocturnal

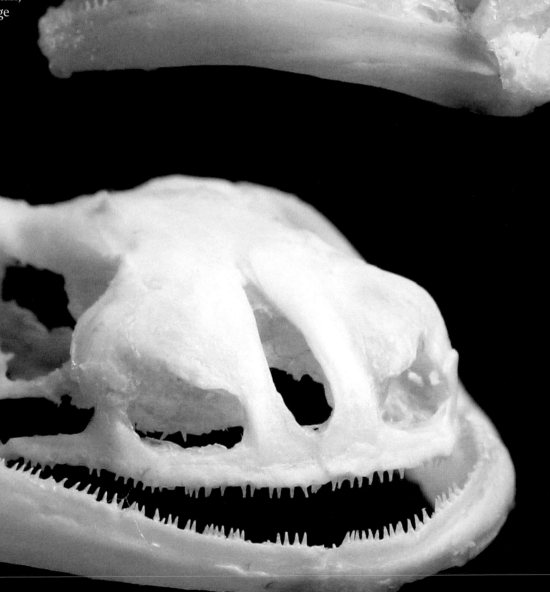

Smooth Newt
Lissotriton vulgaris

There is not a lot to this smooth newt skull—just a jaw, nostrils, orbits, and rows of tiny teeth. This newt is common throughout much of northern Europe, growing from tadpole to a tailed, four-legged beast during late spring and early summer. It is the tiniest skull in Dudley's collection, smaller than a fingernail.

Kingdom: Animalia
Phylum: Chordata
Class: Amphibia
Order: Caudata
Family: Salamandridae
Genus: Triturus
Behavior: Carnivore/Nocturnal

The Nature of Skulls

ONLY A VERY SMALL proportion of the tens of millions of animal species on our planet have skulls: taxonomists estimate a total of just 58,000 or so species. Less than one-twentieth of 1 percent of the entire animal kingdom, in other words, are classified as either vertebrates or craniates—the taxonomic term being virtually interchangeable.

A mere 58,000 creatures—aardvarks to zebras, albatrosses to turkeys, alligators to turtles, codfish to wolffish, axolotls to toads—are recognized as being different from their brethren in possessing backbones, spinal columns... and skulls.

In almost all cases, a skull is made of bone. Bones are living organs, extremely complicated creations that have highly varied looks and feels to them. Moreover, very different weights and strengths and degrees of flexibility are apparent among the bones both within one creature and across the entire realm of creatures (the bones of a sardine, for example, are entirely different from those of a hippopotamus, and the bones of a human head are qualitatively different from those of the human thigh).

All bones, however, enjoy certain common chemical and physical features. They are made in large part of a fibrous protein, collagen, and a bonding form of calcium phosphate, hydroxyapatite, that mineralizes the collagen and gives it enormous rigidity, durability, and strength. The collagen, which is the organic, living part of the bone, gives this very dense and apparently nonliving material the ability to grow, to bend, and, in those occasional cases when it breaks, to heal. Though the chemical and organic composition is the same throughout the bone, the structure is not: the outer part is made up of compact bone, the inner part of spongy bone. It is largely within this inner, spongy part that we see the evidently living part of the bone: deep within a bone is very clearly a living tissue with a blood supply and nerves, the bone's marrow; it generates and keeps the body beyond the bone liberally supplied with the blood that is crucial to all vertebrate life.

There are five recognizably different kinds of bones in the human body: small, what are called sesamoid bones, like the kneecap; irregularly-shaped bones, like the hips and the bones of the spinal column; short bones, like those in the ankle and the wrist; long bones, like those of the arms and the legs; and flat bones—sandwiches of compact bone around a central layer of spongy bone. Flat bones comprise the sternum, or chest bone, and they make up most of the bones of the skull.

In essence a skull—of no matter what animal—consists of two basic, separate, parts held together only by muscles and cartilage. The larger part, usually, is the *cranium*, which holds the brain and offers housing for the sense organs; the other is the *mandible*, more familiarly known as the jaw.

The cranium, like the mandible, is generally equipped with teeth, the leveraged use of which allows an animal to seize and crush and make digestible its food. To the mandible there is little more than this: it's a powerfully muscled crushing device, but little else, except that it occasionally sports some decorative flourishes.

The cranium, however, presents itself in a variety of forms. It has three basic components (although it has very many individual parts, and very many discrete bones). It has a braincase, the often-domed structure whose very name signals its role: to cover and protect the brain. It has what are known as zygomatic arches, bones that provide both the openings and protection for the eyes; and it has structures, occasionally quite spectacular in appearance, that connect the braincase to the third part of the skull, the rostrum. The *Oxford English Dictionary* offers etymologies for all the skull parts, naturally; its entry for the origins of *rostrum* happily gives just about all the forms in which skulls' rostrums present themselves to the world, depending on the animals that own them: so the rostrum can be a snout, a muzzle, a beak, a bill—or a face.

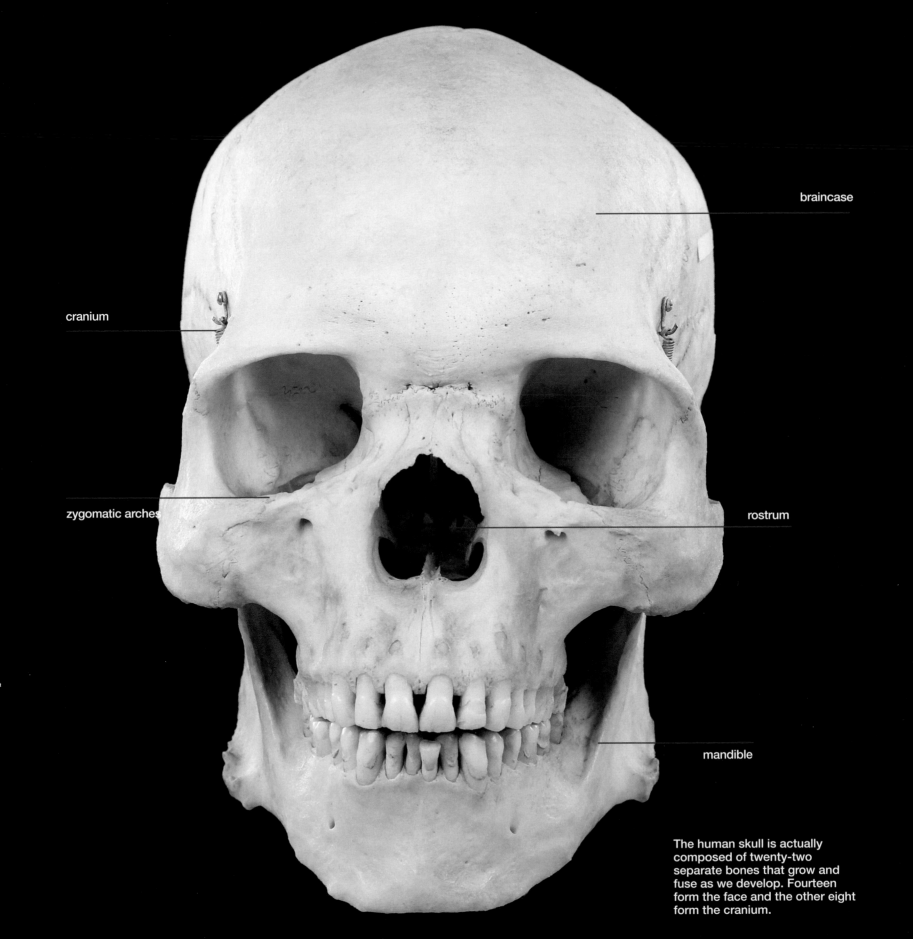

braincase

cranium

zygomatic arches

rostrum

mandible

The human skull is actually composed of twenty-two separate bones that grow and fuse as we develop. Fourteen form the face and the other eight form the cranium.

A Skull's Component Parts

SKULLS MAY LONG HAVE FASCINATED and captivated us, but we have thus far not managed to give more than a very few of their parts the kind of names that might ever become familiar. We recognize jaw, palate, and eye socket, and all would agree that braincase, though less familiar, is an easily understood word. But these few aside, the rest of a skull's construction is a maze of classical names, most of them Latin. English words such as thigh, hip, shoulder, finger, and flange, for individual components or features of other bone systems, have no skull equivalents; we have, instead, the interparietal and squamosal bones; the temporal fossa, which describes an indentation that extends across several bones; and a score of other terms that include one for a tiny protrusion near the front of the palate known as the vomer. The object with so unusual a name is defined as a small, thin bone forming the posterior part of the partition between the nostrils in man and most vertebrate animals, and the word comes from the Latin for plowshare.

These names matter, because it is by the dimensions and appearance of its various parts that a skull is generally classified and described; inevitably, in the descriptive parts of this book, references to these component parts will appear. The condensed version of a skull description (in this case, of the northern river otter, also known as the North American river otter, *Lontra canadensis*) can be simple enough: the skull is flattened and rounded with wide-spreading slender zygomatic arches, very short rostrums, and very large braincases.

But the fuller version can be very daunting: the rostrum (of the aforesaid northern river otter) is very short and broad; the nasal sutures are visible only in immature specimens. The zygomatic arches are very wide-spreading and slender, converging to the anterior; the oblong infraorbital foramina are large. Postorbital processes are well developed and sharp; the orbits are medium, forward-facing, and less than half the size of the massive temporal fossae. The interorbital region is broad, and the interorbital breadth is wider than the very constricted postorbital breadth. The braincase is long, wide, and much more than 50 percent of the greatest length of the skull; temporal ridges are poorly developed and converge high on the skull in adult specimens to form a low sagittal crest. Occipital crests are well developed. The mastoid processes are large and distinctive...

The full description continues for yet another twenty lines, complemented by what looks like a complex numerical formula, which in the case of our otter is written as i3/3 c1/1 p4/3 m1/2.

This formula, crucial to the description and classification of any skull, has all to do with teeth — which ardent skull collectors do their level best to preserve and retain in situ. The four letters in this formula, *i*, *c*, *p*, and *m*, refer to the animal's incisor, canine, premolar, and molar teeth; the numbers give the count of each tooth type on the upper and lower parts of each side of the jaw. Thus, our otter here possesses on each side four premolars on its upper jaw and three on its lower, the mandible. The total number of teeth on each side of the skull is eighteen, and so the total in the otter's head is thirty-six.

Cranium Modifications

THE BRAINCASE OF A SKULL may well be, as advertised, a strongly built and cleverly engineered structure, but listening to all that

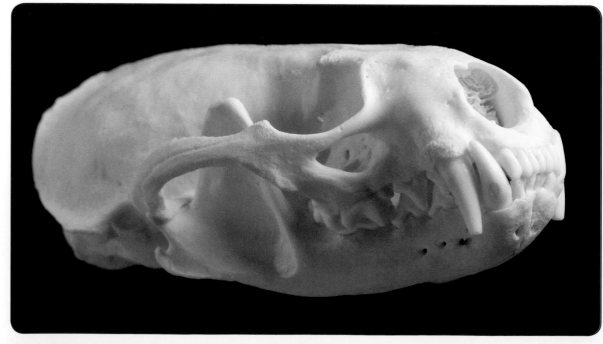

This European otter skull is similar to that of the northern river otter. Although the jaw is closed, it is still possible to identify and count the teeth on each side of the jaw. There are three incisors, upper and lower; one canine, upper and lower; four premolars upper and three lower; and one molar, upper and two lower.

incessant banging coming from the direction of the crab apple tree in the garden, one has to wonder: is it really strong enough to keep a woodpecker from the most terrible headache?

And what about those rams and deer you see butting heads with such determined ferocity during springtime? The echoing, crashing sounds, the visions of extreme and repeated violence, the frequent tangling of horns—how do animals with an instinctive need for such brutish behavior prevent their brains from turning into rice pudding?

(Having said which, it sometimes does seem to the lay observer that rams' brains are pretty much made of rice pudding anyway. A pair will be grazing peacefully; both then look up for no apparent reason, trot toward one another, start smashing furiously into each other's heads for a short while, then back away and stand quietly gazing into space with expressions of vacant stupidity and when done, resume grazing. You half expect them to ask: Now, why did we do that?)

Both the ram and the woodpecker happen to have very dense skulls, especially in that rounded rear area known as the braincase, where they are built like armored cars. Crucially, their braincases are also unusually smooth inside (compared to those of other animals).

The brains of most animals that are prone to head banging—these include deer and other antlered mammals, too, as well as the various birds—are relatively small and (unlike a human's) smooth-surfaced; and they're bathed in only small amounts of cerebrospinal fluid, leaving little room for the brain to move and be shocked by the sudden decelerations and accelerations of their weaponized heads. Moreover, both rams and woodpeckers are scrupulous in the precise, single-direction fashion in which they smash their heads into things, whether trees or each other: the aim is such that there's very little side-to-side torsion exerted on the brain, none of the movement that induces whiplash injury and other kinds of damage.

Gannets have solved a similar problem. These magnificent black-and-white seabirds, with wing-spans of as much as six feet, catch fish by spectacular dives into the ocean. Starting from heights of a 100 feet or more, they enter the water at sixty miles an hour and hurtle downwards far beneath the surface, pursuing their chosen fish underwater, like penguins, using their wings to swim.

It's an awesome performance—not least because they are so successful as hunters: they are eagle-eyed (if the avian mixed metaphor may be allowed), and they have, unusually for birds, true binocular vision, which helps lock them on target. If lucky, they will consume the fish while still underwater, only eventually bobbing back to the surface to take off, something they do very clumsily, and resume high-altitude patrol.

However, their fishing success is one thing. Their survival is quite another. To dive into water from 100 feet may not be lethal for a lightweight creature. (One remembers J.B.S. Haldane's famous essay "On Being the Right Size," and his unforgettable declaration, applicable here: You can drop a mouse down a thousand-yard mine shaft; and, on arriving at the bottom, it gets a slight shock and walks away, provided that the ground is fairly soft. A rat is killed, a man is broken, a horse splashes.)

A gannet would neither be broken nor, in the sense that Haldane meant, would it splash. But it would, or should, get a fearful migraine. Yet that doesn't seem to happen. Gannets manage to bob to the surface with all their mental faculties intact, their brains entirely unhurt. And how? Skull modifications, just as with the ram and the woodpecker. In this case, to mitigate the brain-shattering impact of what is, in effect, a collision with a wall of water at sixty miles an hour, air sacs built into the gannet's face act as a cushion; its extremely long and narrow beak helps the bird enter the water with only a very stealthy kind of collision; and it has no nostrils that would allow water to gush inward at sixty miles an hour and do serious damage to the delicate tissues inside. A gannet's skull is built like the nose of a Concorde, strong, delicate, unpierced, and able to be tilted downward on landing but to be held straight ahead when passing at a high rate of speed through the water.

Modifications of the skull are many, all produced by evolution to give each animal maximum advantage in being able to adapt to its environment and its lifestyle niche. Some skulls are narrow and delicate—the gazelle's, say—while others, like that of the lion, are squat and fat and powerful: the owner of one grazes and has the grace to prance away; the owner of the other stays put and waits for chance to bring him the opportunity to crush and to kill.

Invariably the melding of form and function is displayed perfectly in an animal's skull, such that one can usually, and quite easily, deduce the manner in which the animal behaves, or the environment in which it lives, by examining, or even by casually glancing at, the skull it leaves behind.

Teeth, horns, and beaks we'll discuss in a moment; but the jaws—the mandible below, the cranium above—display variations that indicate in an instant the kind of use to which each animal has put them.

Compare the massive jaws of a wolf, for instance, with the more modest arrangements of the mouth of a hare; or look at the great ridges along the uppermost part of the skulls of some species—a mountain lion, for example, which has a sail-like arrangement known as a sagittal crest, and to which the jaw muscles are anchored to give it even greater crushing power. If you come across an otherwise modest animal with a large sagittal crest—a coatimundi, for instance—you'd do well to avoid it, or at least to keep it happy: for when a coati fights, it bites; and when it bites, the muscles attached to its sagittal crest allow it to come down hard with its canine teeth, the ones that really hurt and do damage.

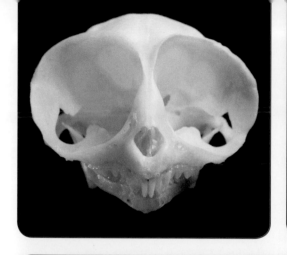

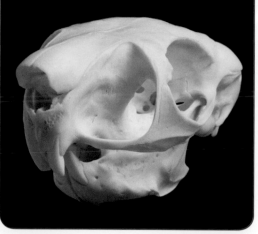

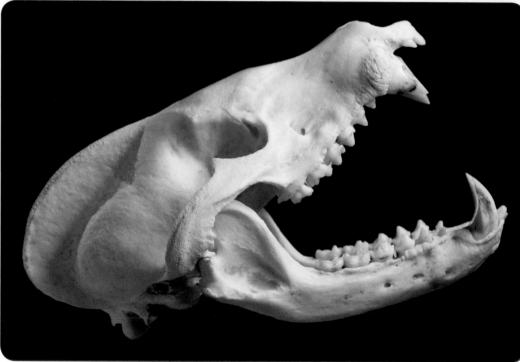

The eye orbits of this Philippine tarsier (*left*), like those of many nocturnal primates, occupy a significant portion of the cranium.

The auditory bullae of the springhare (*right*) are bulbous structures found below the rear part of the skull of placental mammals. They enclose parts of the middle and inner ear.

Carnivores, with their big and powerful jaws, often have pronounced sagittal crests like that of this male coatimundi skull (*below*).

than usual. The rabbit skull, however, is awash with clues as to how this particular animal behaves—which makes it, to collectors, a rather more intellectually satisfying skull to collect, if perhaps not as spectacular to have on display.

Horns

"And seven priests shall bear before the Ark seven trumpets of rams' horns: and the seventh day ye shall compass the city seven times, and the priests shall blow with the trumpets."
—Book of Joshua, Chapter 6, Verse 4

STUDENTS OF THE BIBLE will immediately recognize the event: the destruction of the city of Jericho, the first skirmish in the Israelites' attempts to conquer the land of Canaan. But biologists, and those perhaps less concerned with mystical Holy Land history, will find in these biblical lines something more intriguing: the identification of the musical instrument that was chosen by Joshua to warn Jericho's inhabitants that their city walls were about to fall. It was the shofar, the ceremonial trumpet fashioned from a bony excrescence on the head of a male sheep—a ram's horn.

Indeed, it can hardly escape notice that today's musical instrument shares its name with the name of that excrescence—the word *horn* being common to both. It is a word with both Greek and Latin origins and yet one that, when it comes to English, around AD 825, has a rather intriguing history.

(Be doubly wary if your coati is male; its sagittal crest is larger than those of females. Sexually dimorphic arrangements of sagittal crests are apparent in fisher cats, too—so beware, as well, of angering a male fisher cat.)

In less threatening territory, the eye orbits on certain skulls can be spectacular—the immense orbits of a tarsier, for example, are often as large as the entire rest of the skull; they provide a classic example of a skull detail that suggests how well or ill an animal can use a certain sense, in this case, its vision.

Similarly, what are called the auditory bullae can show, in bony form, just how well an animal can hear. Springhares and

rabbits have very large auditory bullae: it's said that the chambers enclosed by the bullae resonate perfectly to the whooshing sound of a downward-swooping owl, enabling the rabbit to dive for its burrow and live to enjoy another day.

Until its skull was examined, no one had any idea of how a rabbit was able to do such a thing: the owl flew down, the rabbit vanished—such magic long seemed as mysterious as the butting of rams and the vacant stares that follow.

As always when at a loss, animal behaviorists like to say that the rams carry out their butting for reasons of power, territory, or sexual ritual. Their skulls give nothing away, other than being stronger

In its first English appearance, it describes both a musical instrument and a biblical symbol of power. But then, almost two centuries later (around AD 1000), it appears also to signify both the specific animal excrescence and something more utilitarian, a container for powder, oil, or ink, made from the excrescence after it is separated from the animal.

At first this seems a puzzle. Why does the instrument come before the animal part? That sequence seems to imply that an animal's horn (the later English sense of the word) is shaped like the horn-as-musical-instrument (the word's earlier usage). But that, of course, is quite absurd, counterintuitive, and wrong. For the reverse has to be true: the musical instrument (a modern invention) has to be shaped like the animal part (something that would have been on show for mankind since the most primitive times). So why the strange order of appearance of the English word *horn* in these two most basic meanings?

The answer has to be that the English language can occasionally be very slow in assimilating certain concepts, for reasons that are linguistically rather (dare one say it?) snobbish. A thousand years ago English people—learned English-speaking people, that is, the types who were given to writing about parts of animals—were in all probability still quite content to employ the Latin word—*cornu*—when dealing with skeletal anatomy, even though they had already been using for some two centuries the English word—*horn*—to designate both the musical instrument that looked like a cornu and the biblical instrument of power that actually was made from a cornu. (The word *cornet*, which signifies a small trumpet to this day, is a much more direct descendant from the Latin.)

The Bible is littered with uses of the word *horn*: there's a bullock with horns (Psalms 69), a beast with ten of them (Daniel 7), and a lamb with seven (Revelations 5), and references to horns of iron (in 1 Kings 22) and (in Micah 4) to a metallically super-equipped animal with not only horns of iron but hooves of brass.

The horn of an animal, in short, has long been regarded as important and, at least in the Holy Land, highly symbolic, nowhere more perhaps than among the Jews, whose use of the ceremonial horn, the shofar, plays a central part in innumerable religious ceremonies. (Alexander Cruden, the exotic Scotsman who compiled a famous early concordance to the Bible that lists every word in the Good Book and where to find it, helpfully concludes his lengthy entry on horn with "See Ram's".)

Thus to the shofar. Back in rabbinical history a shofar could be long or short, plain or adorned (with a gold or silver mouthpiece, for instance), straight or curved, fashioned from the horn of a sheep (for ordinary, daily blowings in the temple, to make ordinary announcements) or specifically from the horn of a ram (for those major anniversaries and congregational summonses), or from the horn of a kudu and serving both functions (in the case of Yemeni Jews). Shofars were used by soldiers as a call to battle—as in the demolition of the walls of Jericho. The shofar is still used once in a while to make a political statement: at the end of the 1967 war, Israeli rabbis blew the shofar at the Western Wall, in celebration.

As with most aspects of Jewish life, strict rules attend to the making and use of the instrument. A shofar must only be made from a kosher animal—from a member of the Bovidae, a cloven-hoofed animal that chews its cud. So the horns of bison, antelope, gazelles, sheep, and goats are in theory allowed, while a shofar can never be made from the antlers of a deer, because these are not horns but are made entirely of bone and cannot easily be hollowed. A horn, though it often has a core of bone, is made principally from alpha-keratins, the same proteins found in fingernails and hair, and so is much easier to work.

Today the use of the shofar—and in modern practice almost all the best shofars are made from the curved horn of a ram—is generally restricted to the celebrations of Rosh Hashanah and Yom Kippur. The person selected by each synagogue to play it, blowing a sequence of notes that are intended to sound like the voice of God, is deemed one of the most honored in the congregation.

Horns are part and parcel of most male bovid skulls—skulls that themselves are modified to accommodate what these cow-like and cloven-hoofed creatures do for most of their lives: graze. So the teeth are flat and sharp; the eye orbits are large and set well back into the head so that the animal can view the world even though its head is bent down into the grass; and the upper cranium is reinforced, with sinuses and other well-engineered strengtheners, to allow it to indulge in the mating and territorial behavior rituals of head butting that are, or were, so central a part of its character. It is thought that horns originated as an adaptation to this behavior, an outgrowth from the skull that further protected it from butting damage: the bony core of a horn is organically identical to the skull bones from which it grows. As horns evolved further, they reached beyond protection to the point where they themselves can inflict damage on an enemy or a fellow bovid who might wish to engage in competitive butting.

Today most shofars are made from rams' horns and are blown in synagogues on important holidays.

Horns can be used to tell the age of an animal: they are ever-growing, doing so in measurable steps that can be read just as are tree rings, and often perform interesting osteogymnastics as they do so—extending (think of domestic cattle) horizontally outwards or diagonally upwards at first, then sometimes curving straight up (think Texas longhorn or Asian zebu), or backward (bison and wildebeest) or becoming massively recurving spectaculars, as in mature and (sadly) eminently huntable mouflon sheep.

Perhaps the loveliest and most graceful of all horns are to be found on an extremely rare southern African antelope, the giant sable antelope, *Hippotragus niger variani*, the national animal of Angola that was—somewhat ironically—rendered almost extinct by that country's thirty-year-long civil war, which ended only in 2002. The huge, swept-back horns of the male of the species frequently reach lengths of six feet. This shy, immensely powerful animal, when standing erect in the forest displaying its great horns, is one of the more striking images of animal grace and dignity, especially now that it's so seldom seen.

Bovids have horns with a single shaft. On the heads of some non-bovid ungulates, such as the pronghorns, a non-antelope antelope-look-alike found in North America, the bony core of the horn splits (and since most have bony cores, the split extends back into the skulls also), leading to the development of doubled horns that in appearance are a little like a deer's antlers (which, to reiterate, are not horns at all, but uncovered bones that drop out each year and regrow). One of the tines of a pronghorn grows forward, as a prong.

Bovids and pronghorns are part of a larger group of even-toed ungulates (which bear their weight on the fourth toe, usually) that also includes pigs, peccaries, camels, mouse deer, cattle, goat, sheep, hippos, antelope proper (which do have horns—the blackbuck having spectacular three- and four-turned spiral horns), and giraffes (which have vestigial, fur-covered, bony bumps on their heads, called ossicones).

There are odd-toed ungulates, too, whose weight is borne by the very much larger middle toe on each hoof; they include zebras, tapirs, and, crucially for horn fanciers, rhinoceroses, which, more than all other creatures that sport these bony excrescences, are threatened with extinction precisely because they have them.

Blackbuck antelopes are hunted for their skin and meat, and bighorn sheep are hunted so the entire head can be mounted on some vanity wall, but the five extant species of rhinoceros—two in Africa and three in Asia—are hunted quite specifically, and dangerously, for their horns.

Rhino horns are different from most others, in that they lack a bony core, consisting exclusively of what other horns possess only as their outer covering—the keratin that elsewhere forms fingernails, toenails, and hair. The Indian and Javan rhinos have a single horn; the two African and Sumatran species have a pair, one in front of the other, as much as five feet long. All rhinos are equally hunted, mostly by poachers eager to sell the sawn-off horns on the black market—either for use as trophies (dagger handles, mostly, in places like Yemen) or, more frequently, to be crushed and ground up and used in traditional Chinese medicine as curatives or aphrodisiacs (along with tiger penises, snake gall-bladders, and other improbable anatomical oddities). Although some practitioners of traditional Chinese medicine continue to claim spectacular results for powdered rhino horn in limiting convulsions and lowering fevers in human patients, China itself has signed a treaty banning the trade in rhino parts, and there is hope that the populations of the endangered sub-species (such as the northern white, of which only four were known in 1998 to exist in the wild and which now number fewer than a dozen, all in captivity) may begin to recover.

There are horns to be found on other, non-mammalian, creatures too: the appropriately named rhino iguana as well as some varieties of horned lizards (Wyoming's official state reptile) whose keratinoid growths are bony-cored and are thus true horns.

There's something of a parable here. For, as with the rhino in the medicine cabinet and the bighorn sheep on the trophy wall and the ram's horns in the synagogue, so the mystery and allure of horns in all animals seems to have a transcendental role for the human race, wherever in the world these humans live, and however majestic or prosaic the animal that sports the horns happens to be.

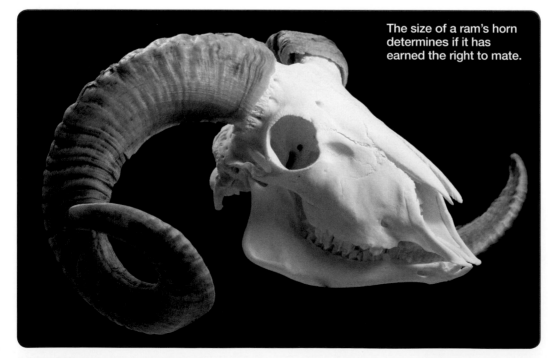

The size of a ram's horn determines if it has earned the right to mate.

(Clockwise) Giraffes have unique horns that may number from two to five across the top of the head.

Deer grow an impressive set of bony antlers that they shed each year. Horns grow much more slowly and never shed.

The distinctive horns of the black-buck can spiral around up to four times.

Pound for pound, rhino horn is more expensive than gold or cocaine. This example from the Field Museum in Chicago may be worth more than $400,000.

This Sulawesi wild pig's skull has impressive tusks.

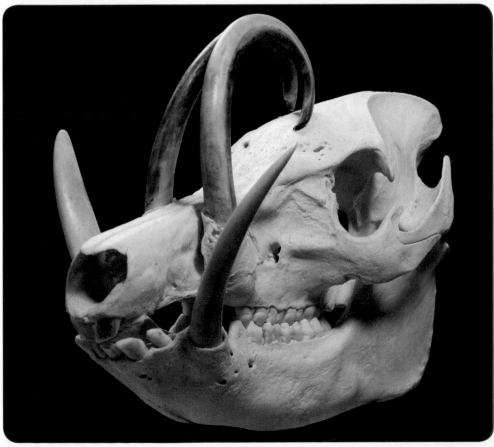

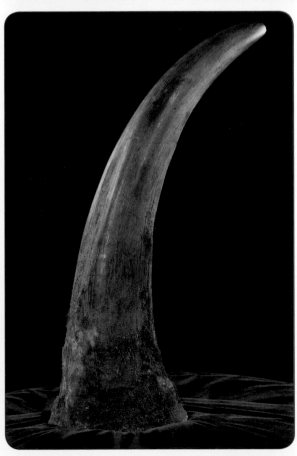

Teeth

There was a young lady from Niger
Who smiled as she rode on a Tiger
They returned from the ride
With the lady inside
And the smile on the face of the Tiger

MAKE AN EIGHT-TOOTH smile: such is one of the cardinal instructions handed down to the young ladies who currently work on China's new fleet of high-speed trains. To help them in this endeavor, they are given exercises that involve chomping down on chopsticks, sideways, for hours at a time. The result—the prominent display of two rows of perfect, titanium-dioxide-white teeth framed by two perfectly shaped and carefully parted pink lips—is regarded in today's China, and indeed among most of humankind (though not in Japan and Vietnam, among others), as indicating the highest degree of empathy and amiability.

A full-on smile signals friendliness, happiness, approachability, and good humor. And while it is entirely possible to smile without showing the teeth, it appears axiomatic in what is known as the hospitality industry, as well as in Hollywood and on television, that such a half-hearted display is almost unthinkable: the show of white teeth, and lots of them, is near-universally a joy to behold.

But this is among humans. In almost all other animals, the baring of teeth—even the pretty pearly white teeth of Bertolt Brecht's *Threepenny Opera* shark—suggests something very different. You run towards a human being who flashes you an eight-tooth smile. But when an animal does it you run the other way, as fast as you can. Teeth, generally speaking, are scary.

By chance—and it is just chance; the etymology of the word is quite different—the animal that most dramatically displays a scary animal-smile is the giant extinct beast known as the smilodon, also (but incorrectly) known as the saber-toothed tiger. Technically, it wasn't a tiger, just a cat; but it was a cat with spectacularly huge teeth—a pair of them, downward pointing. Curiously, these formidable dental items

were not used as weapons, not to bite into an animal to kill it (the smilodon had a relatively gentle bite). Instead they were used to finish the creature off once the smilodon had subdued it with its well-muscled forepaws and vicious claws.

Of today's animals, the rare clouded leopard has perhaps the most wicked-looking teeth. Its bite is powerful, and it does hunt and uses its jaws as a weapon. Red in tooth and claw may be a phrase that isn't entirely applicable to the saber-toothed cat, but it works for almost all of the extant felids, lions to cheetahs to tigers and leopards. If they smile, you run.

And the teeth of these great cats from which you run are—not always, but mainly—those usually large and sharp ones close to the front of the mouth, the canines. These

are the teeth of the midnight movies—teeth annealed onto the powerful side bones of the face, and purpose-built to cut and puncture and slash and pin the flesh of enemies or prey.

Covered by an incredibly hard crystalline form of calcium phosphate, known as enamel (beneath which in many animals is dentine, or ivory), teeth survive long after the body with which they are associated has decayed into dust. It's common that, when a skull is found, the teeth are found in it or close beside it. Indeed, teeth will usually last longer than the skulls in which they are embedded. Of those teeth, the canines are usually the most prominent, their size and shape often offering an instant guide to identification of their owner.

Human skulls certainly do display canine teeth—four of them, the third tooth from the front along on each side,

The teeth that sit in the impressive jaws of this saber-toothed cat are, alas, not real.
The skull itself is also a replica.

above (in the cranium) as well as below (in the mandible)—but in appearance the difference between human canines and their neighboring teeth is hardly dramatic. True, they are conical and pointed, while the incisors in front of them are chisel flat and the premolars and molars behind them have generously rounded cusps for grinding. But human canines are not very much larger than the rest—except, of course, when the human has become a vampire, in which case the canine teeth are ludicrously elongated, the better for sinking into a victim's carotid artery. Other than in horror fiction, humans do not employ their canine teeth for fighting; and if ever a human bites another, it's usually with the front, incisor, teeth.

It's in the carnivorous mammals—the bears, the big cats, dogs, peccaries, and pigs—and in some of the larger reptiles (such as crocodiles, whose "canines" are actually not true canines) that the canines (hence their name, from the Latin for dog) are most developed, and are most terrifying. The afore-mentioned saber-toothed cat, for instance, had upper canine teeth that could be as much as two feet long, extending well below its head even when its mouth was closed. Even on a smaller scale, the overlapping nature of the upper and lower canines can be alarmingly distinctive: a flat piece of paper placed between the closed jaws of a tiger, a wolf, or a polar bear would be pierced in only four places—where the canines drive past one another, the upper pair pressing deep into the outer mandible, the lower ones pushing high up towards the side of the face.

In some other animals—animals that are seldom carnivorous, in fact—the canine teeth grow continuously (unlike most teeth) and are even more elongated, arranged in curves or helices or squashed circles and generally known not as teeth but as tusks.

A warthog's tusk teeth, wickedly curved, are used (besides as fighting instruments) for digging. Similarly, the very prominent upper pair of tusks of a walrus help the animal haul itself out of the sea onto the ice, to help move its blubbery body along the ice, and to dig holes on the shallow ocean floor to gain access to its prey.

But not all tusks are canines. A narwhal has a single tusk—one that invariably, and oddly (considering the bilateral symmetry of most chordate creatures), grows out of only the left side of its jaw—that is a very large incisor tooth. Elephants, too, sport incisors that are elongated and often curved—the most famous tusks of all, of course. Indeed, elephants are unusual in having no canines. And, as a further reminder of the imperfect nature of symmetry, their tusks are of

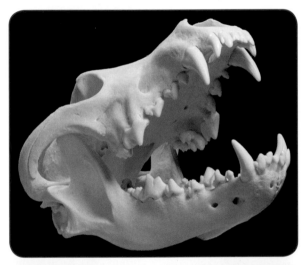

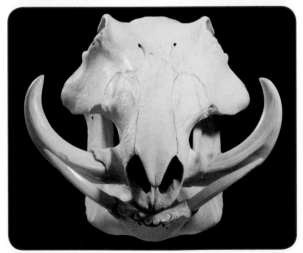

(Clockwise) Members of the dog family, like this gray wolf, have dentition similar to a cat's.

The Nile crocodile's smile is truly terrifying.

Rodents' incisors are usually harder on one side than on the other allowing the upper and lower sets to wear against each other and self-sharpen like chisels.

Its self-sharpening tusks help the warthog defend itself against predators. They also enable it to dig underground dens for shelter.

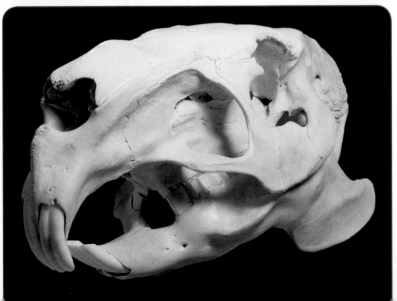

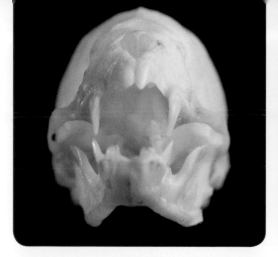

The teeth of the infamous vampire bat (*left*) are used to shave the bite region of hair before inflicting their painless wound.

Gaboon vipers, like this one (*right*), have some of the largest fangs in the snake world and should a pair become damaged, there is another set waiting just behind to replace them.

different sizes: like humans, elephants are usually right- or left-tusked.

Teeth are hugely important in suggesting evolutionary origins. The fact that elephant teeth have much in common with those of the manatee supports the persistent belief among evolutionists that elephant history may have included an aquatic phase.

Since for thousands of years a vigorous trade in elephant ivory has flourished (now greatly reduced, under terms of CITES, because of the danger that ivory-hunting presents to elephant populations), it's widely assumed that only a few animals have teeth made of ivory. But in fact almost all teeth (including those of humans) are made of it. No one, however, hunts primates of any kind for ivory—because other than elephant, walrus, and narwhal, few teeth are prized enough, or are large enough, to be used in jewelry or furniture or piano keys or for a seaman to practice scrimshaw. Moreover, the enamel that covers ordinary teeth renders them a great deal harder to work.

Incisor teeth are seldom as pronounced as canines—though being at the very front of the mouth, they are particularly visible: when a skull grins at you, it does so with its incisors. The animals that have the most distinctive incisors are, of course, the rodents—from the Latin, animals that gnaw. Gophers and beavers, and the magnificently large and lazy capybaras, all have highly developed incisors, caricatured in countless cartoon films. Nipping, scraping, de-barking, fur grooming—these and many more are the functions performed by the incisor teeth of rodents, which can be incredibly sharp, incredibly strong, and seemingly ever-lasting (in fact, many keep on growing as fast as

they are worn down by work, and are self-sharpened by the constant gnawing and tearing).

Horses and cows, too, have powerful and sharp incisor teeth, used—in combination with the tongue—for ripping up grass and hay and shoving it to the back of the mouth to begin the process of digestion.

Vampire bats' incisors have astonishingly sharp edges. They lack the enamel that blunts (and preserves) most other teeth, and are used to inflict the tiny cuts, just a few millimeters long, that are painlessly made on a host animal's flank; the bat then trickles anti-coagulant-laced saliva into the wound, which enables the victim's blood to flow and be swallowed by the vampire until it weighs more than double its usual weight and has to rocket-propel itself upward in order to gain the lift needed to fly home and begin digestion.

In contrast to the sharp incisors of bats, snakes have either grooved or tubular incisors down which, or through which, they may inject venom into victims or enemies. (They also often have backward-pointing teeth behind the incisors, enabling the snake to slide its mouth onto its victim's limb, like a glove—and not release until the mouth is opened wide.) The Gaboon viper, an especially lethal six-foot-long specimen found in the jungles of western equatorial Africa, has a pair of fangs that are two inches

long. A modified salivary gland near the base of these teeth produces and stores large quantities of venom when the snake strikes, it holds onto its victim and closes its jaws hard, so that the poison from the sac inside its mouth is squeezed down the two-inch-long tubes and deep into the victim's musculature.

Some cobras have a further modification: a venom channel that makes a sudden sharp turn towards the end of the incisor tooth and enables the snake to spit its poison, often as far as six feet. It generally aims for the eyes; the accuracy of its spitting, and the efficacy of its venom have caused numberless cases of blindness in areas where this breed of cobra is common, mostly in Somalia, Uganda, and northern Kenya.

Not all venomous snakes have tubular incisors: most have a groove into which the venom is pressed from a nearby gland. In this respect some snakes are similar to the very small number of mammals that are venomous. Among the most primitive mammals are the shrews, some of which have poisonous saliva. One kind of large shrew, the solenodon, peculiar to Cuba, has a grooved lower incisor through which it delivers its poison.

Alligators and crocodiles have formidable arrays of teeth, all of which are constantly replaced as they grow out—a crocodile can exchange fully 3,000 teeth during its lifetime. (A handy tip: you can usually see a crocodile's lower canine-like teeth when it has its mouth closed, whereas all of an alligator's teeth are hidden in sockets when it shuts its jaws—an easy way to tell one animal from the other.)

It's widely assumed that only a few animals have teeth made of ivory. But in fact almost all teeth (including those of humans) are made of it.

A bird's beak, like this toucan's, is primarily bone covered in a layer of keratin over a vascular layer containing blood vessels and nerve endings.

Numerically, the molar and premolar teeth usually form the bulk of a mammal's dentition: they are the teeth that do the hard graft once the weaponry of the front teeth has done its work. When the prey has been caught or the enemy killed or the grass shorn from the field, the molars begin the grinding and shearing and crushing that reduces the food to digestible size and form. Saliva—strong, and often highly acidic, digestive juice—is added to the mix, and the resulting slurry is sent back into the esophagus and on to the stomach to be converted into useful metabolic fuel.

The design of molars varies according to the kind of food the animal consumes: sea otters have teeth that will crack shells, wolves can crush bones, bears use theirs for smashing down plant materials, moles have pointed back teeth to crush the hard exoskeletons of insects, dolphins have conical molars for holding and repositioning slippery fish so that they can be gulped down into the throat. Molars may not be the most dashingly romantic of teeth, but, like superb pieces of massive Victorian engineering, they are fit for their purpose, and impressive by being so. It is this specialization in dentition (what the scientists call heterodont), and the associated skull structure, that has led to the extraordinary adaptive success of the mammal group.

Finally, what of the shark, with what Bertolt Brecht called its pretty, pearly whites? The teeth of the most vicious of sharks are truly terrifying—in their shapes, in their sheer numbers, and in their cunningly

deadly arrangement. Row upon row upon row of razor-sharp, often-serrated, usually triangular, very hard, backward-pointing, and lethally arranged shark teeth are discarded and then replaced with great rapidity—a shark in its lifetime may own 35,000 of these horrors, using them to deadly effect on other swimming creatures, including, if the opportunity presents itself, the limbs (or worse) of any human incautious enough to pass close by. Bertolt Brecht wrote of them sardonically; Peter Benchley, in *Jaws*, wrote of them more realistically. Few who ever saw that classic film of the mid-1970s will ever forget just how awe-inspiring is a great white shark, how formidable its armory of hundreds upon hundreds of pretty, pearly white, and well-nigh indestructible teeth.

But (and it's a big and important but) for all the importance of their teeth-filled mouths in popular scare culture, sharks do not have bony skulls. Tigers do. Wolves do. Crocodiles do. But sharks do not. Sharks do not have skulls, in the real sense. They have a tooth-filled front end, true—but this front end is made of cartilage, not bone. So sharks are in this respect quite different to all of those other well-muscled and dentally dangerous predators of which we are all so rightly scared. And because of this, because of their difference, they're not here.

So you can have all the nightmares about them that you like, but we don't find actual examples of them in Alan Dudley's collection; and for the most pedantic reasons of biological exactitude, you won't find pictures of them in this publication either.

And yet they are so culturally

unforgettable that we couldn't quite ignore them, so our way of including them has not been a matter of taxonomy and the complex subtleties of biology, but rather by way of reminding you of the *Threepenny Opera* song and of all of the sleepless nights that thoughts of these fearsome creatures may well bring to you.

Beaks and Bills

DOROTHY L. SAYERS, one of Britain's greatest detective story writers—she created the Lord Peter Wimsey series, among other triumphs—was in her youth a copywriter in a London advertising agency. She was quite brilliant at her job: within months of arriving she had created for the Dublin makers of the venerable velvet stout the slogan "My goodness, My Guinness," which lasted on billboards well into modern times.

And then one day in 1924 she had a sudden new idea. She drew a hurried sketch of a bird with a remarkable appearance and underneath it then scribbled the draft of what remains one of the best-loved of all British advertising jingles:

Toucans in their nests agree
Guinness is good for you
Try some today and see
What one or toucan do.

The toucan has a particularly fine version of a feature that essentially all birds, and a scattering of other animals, possess: a skull structure (in the toucan's case, one that is colorful, gracefully curved, and unforgettably caricaturable) that encompasses, among other functions, those that elsewhere are usually devolved to the mouth and the nose, a structure called a beak or, in some circumstances, a bill.

A beak can be a tool that offers a bird a range of abilities: it can be a tool for feeding the young; it can be used for applying avian cosmetics or for ruffling

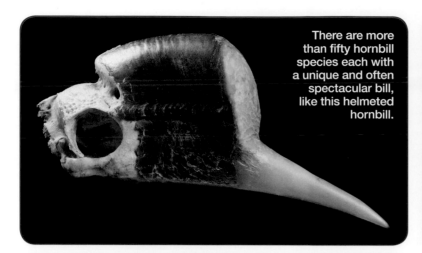

There are more than fifty hornbill species each with a unique and often spectacular bill, like this helmeted hornbill.

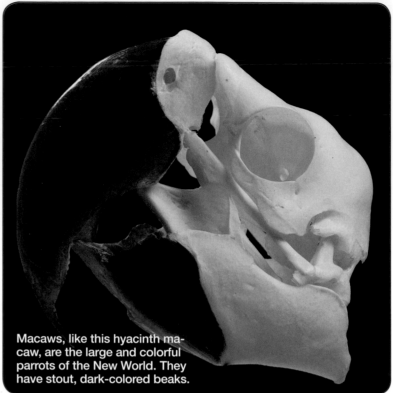

Macaws, like this hyacinth macaw, are the large and colorful parrots of the New World. They have stout, dark-colored beaks.

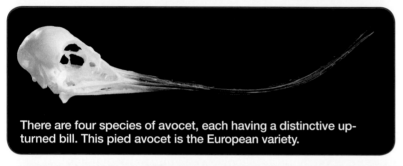

There are four species of avocet, each having a distinctive up-turned bill. This pied avocet is the European variety.

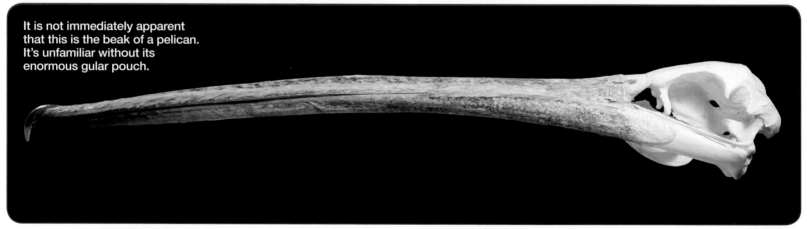

It is not immediately apparent that this is the beak of a pelican. It's unfamiliar without its enormous gular pouch.

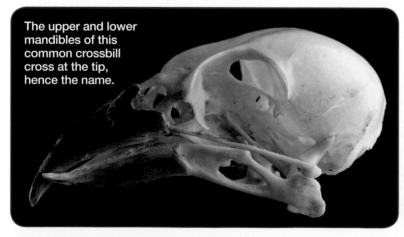

The upper and lower mandibles of this common crossbill cross at the tip, hence the name.

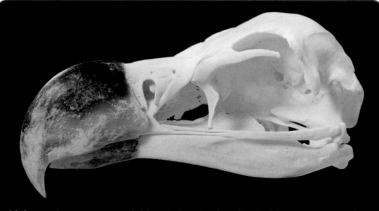

Vultures have a powerful beak for ripping flesh. Many species also lack head and neck plumage, which helps keep the bird's head clean when it feeds from a carcass.

and cleaning the feathers; it can be used for searching out and carrying away prey or (if it's sharp) for killing other animals. A bird can smell through it, breathe through it, and eat through it. If the beak is colorful and grand, the bird may use it to impress and attract the opposite sex, for courtship rituals and mating dances.

A beak is, in other words, like a bird-mounted Leatherman, a multifunctional Swiss army knife bolted onto the front of the skull and protruding dramatically through the feathers of the face—especially dramatically in a bird like the toucan, which has a beak with a truly memorable, quite absurd appearance.

The beak of a toucan is usually about half the length of its body, a great, downward curved, banana-like structure, often yellow, with tiny forward-pointing serrations on the larger upper part (known, in all beaks, as the maxilla) as well as on the smaller lower part (called the mandible). The beak looks so big that the bird appears top-heavy, as though it might topple forward, especially when it perches on a tropical tree branch in one of its jungle habitats. The toucan's small, round eyes, immediately behind this beak, often give it an alarmed look, as if it knows it's about to topple, or else how ridiculous it looks, and hopes (in either case) that no one notices. It can hardly be said that the toucan is proud-looking, though it wears its appearance with a kind of forced dignity.

The serrations led early observers to suspect that the toucan caught and ate fish or was a carnivore; we now know it eats only fruit. In fact, it is so little a hunter-gatherer that its most common stance is at rest, in one place on a fruit-heavy tree, from which it now and again reaches out, using as little energy as possible, to pluck an item to eat.

The discovery that the toucan is basically frugivorous—though it will occasionally eat an insect or two, for variety, and even a lizard—led the Kellogg Company to choose the bird as a mascot for a particularly disagreeable breakfast cereal known as Froot Loops, "Toucan Sam" began his career sporting a beak with (unlike a real toucan) two pink stripes, and

uttering the slogan "Follow your nose, It always knows"—the presumed assumption being that people who ate the cereal were either too young or too uneducated to realize that a bird's beak is not its nose, but instead its mouth and very much more.

But what more, exactly? First of all, the toucan's amply colored (though seldom pink) beak is not thought to be employed as a flirtation device, since male and female toucans have an almost identical appearance; rather, it appears to be used most often as a tool, and for fairly prosaic tasks: it helps the bird winkle fruit out of difficult-to-reach places in trees; it helps with nest building; it accommodates the toucan's six-inch-long and highly sensitive tongue; and it's said to intimidate smaller birds should the toucan wish to go on a nest-robbing expedition.

The beak's inner structure—a kind of spongy keratinous fiber, in texture a bit like the kapok filling of a life preserver—may also help to keep the bird cool in the long tropical afternoons.

As in all birds, the outer part of the toucan's beak is composed of a strong, light, and shiny keratin sheath known as the rhamphotheca, much akin in material to that of spurs and claws. Beneath this

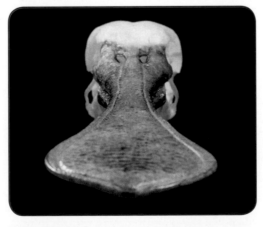

Spoonbills are a group of long-legged wading birds that sift the water in a side-to-side raking motion.

are the blood vessels. Then, a few microns further, come the bony struts that support what otherwise would be a most unwieldy structure; and finally, inside, the kapok-like keratin.

The spongy nature of the beak's filling ensures that, though it may look heavy, it isn't—though it is exceptionally light and strong. Even laden down with berries and pieces of mango and plums and the occasional gecko or mantis, the toucan's beak is never heavy enough to cause the bird to lose its balance and topple over. Without unduly anthropomorphizing, it seems fair to say that the alarmed expression on a toucan's face is just its regular daytime face, the bird behind it quite confident that it can stay upright just as long as it likes.

The variation of types of beak within the general bird population is staggering—every function imaginable appears to be performed with the employment of differently shaped beaks—the tiny bills of grain-eating goldfinches; the filter-equipped beaks of water-sucking birds like flamingos, which are deployed upside down; the massive and arcuate beaks with which spoonbills sift through mud in the shallows they like to inhabit, long, needle-like beaks for nectar-sipping hummingbirds; the long tweezer-like beaks with which a heron catches fish; the huge, curved, toucan-like beaks of a hornbill, with a thick excrescence on the maxilla, known as a casque (often carved elaborately by hunters), which these magnificent birds often use for fighting; the drill-bit-like beaks of woodpeckers and sapsuckers; nut-crushing hooked and very sharp-edged beaks used with great effects by parrots and macaws. Then there are the upturned beaks that enable wading birds like avocets to filter food from the mud; the strong jawed and very sharp beaks of carnivorous raptors; the savage-looking tearing beaks of vultures; the great sac-like lower beaks of birds like pelicans, which scoop and catch and dip net for fish (a wonderful bird is the pelican—its beak can hold more than its belly can); and the apparently harmless-looking, flattened beaks seen on ducks and swans.

Moreover, a good number of these same beaks, no matter what specialized task each may have been modified to undertake, are also used for elaborate male-female bonding rituals, most notably in a process known as billing—as in the phrase billing and cooing, which is now translated, of course, to romantic behavior among young humans. Male and female albatrosses, gannets, and puffins, for example, are especially known for billing: they may clatter their beaks together, they may insert one beak inside another (waxwings are notably keen on this, displaying a kind of avian French kiss), or else they may nibble one another with apparent affection.

But one thing a bird's beak never does do is nibble its food: a bird swallows its food whole, and the break-up and predigestion take place in the gullet, en route to the stomach, not in the mouth-nose arrangement that is enclosed within the beak.

Beaks played a core role in Charles Darwin's nineteenth-century construction of the theory of evolution through natural selection. The small finches on the Galapagos Islands, which he discovered and collected and brought back to London, turned out on inspection not just to be variant forms of the same bird, but entirely different species, each with a different beak design that was suited to one or another of the varieties of ecological niches on the islands. From studying these beaks, Darwin was able to theorize, credibly, that natural variations in their designs among birds that had been born on the islands (all of them, descended from a South American colonist-ancestor) had been selected according to how well each design performed in a particular niche; and that birds with each successful design had bred and flourished, and eventually ceased to breed with birds whose beaks

were of different designs, such that totally new and distinct species had thus emerged. A further study conducted in the mid-1990s on the beaks of what are now known as Darwin's finches has shown that evolution—the process of genetically based selection from within the naturally occurring variations in a population—occurs with a speed that is almost visible in real time. Darwin, in other words, had no idea just how dramatically his theory was put into practice by the birds of the islands where he first came up with his revolutionary (and to some, still controversial) ideas.

One final feature is worth remembering when trying to imagine the look of a bird's beak from the structure of its skull. Beaks are sheathed—they are equipped, as we've mentioned, with a shiny outer covering known as the rhamphotheca. This sheath is not always similar to the shape of the bone keratin within, which is what adheres to the skull. Normally, they do match—sheath mimics bone; one can deduce the look of a bird's beak from the shape of its skull.

But this is often not the case. The red crossbill, for example, has a curved outer sheath (a curved beak, in other words) but a straight bill bone. In the kestrel, there's a notch in the lower edge of the bill sheath, but no notch at all in the bone beneath. It can get quite confusing, and one often has to employ imagination—for beaks can go wild, in ways that the bones beneath them sometimes do not.

A very few nonbird species have facial or skeletal features that look like beaks but are different in form, origin, and function from the sheath-covered bony excrescences that are more properly associated with birds.

There is, for example, the duck-billed platypus , which is a monotreme—an egg-laying mammal, found only in eastern Australia. The platypus has a rubbery device on the front of its head that is more snout than beak. Unlike a bird's beak, which separates to reveal its mouth, the "beak" of the platypus has a mouth in its ventral surface, nasal orifices on its dorsal

(top) surface, and earlike canals just behind the nostrils.

A type of salamander known as a siren, found mainly in the Southeastern United States, has an almost entirely toothless mouth, but with a sheathlike structure on each of its jaws that, taken together, resembles the beak of a bird.

The turtle has a well-defined beak, a protrusion from the cranium that is in fact the animal's upper jaw: there are no teeth, either on the lower edges of the beak or the upper edges of the mandible, but instead bony ridges that serve as cutting or crushing tools. The snapping turtle—with an ill-deserved reputation for ferocity—once saw its beak featured in one of America's first-ever political cartoons: with its name Ograbme—the reverse of embargo, which was the topic of the cartoon—it was shown pulling on the coat of a politician, its beak exaggeratedly prominent.

Then there is the venomous puffer fish—the fugu of James Bond and Japanese cuisine fame—belonging to the family Teraodontidae, the name coming from the four fused teeth that the fish uses for crushing up the shells of mollusks that are its usual prey. Those fused teeth have an uncannily beaklike appearance.

Then we have the parrotfish, which, in a host of colorful coats, grazes algal growth from coral reefs in tropical seas with an ever-growing "beak"—in fact teeth, again—in this case, numerous, tightly packed teeth, arranged on the outside of its jaws.

None of these—platypus, turtle, siren, puffer-, or parrotfish—possesses a true beak, in the sense of the bony structure found protruding from the heads of birds. But that does not prevent all of these creatures, just like the toucan, from being portrayed as being so equipped by the more imaginative of artists, who, like most, have a fascination with the exotic and colorful nature of beaks. Or bills.

Just as, eighty years ago, did Dorothy L. Sayers.

Birds

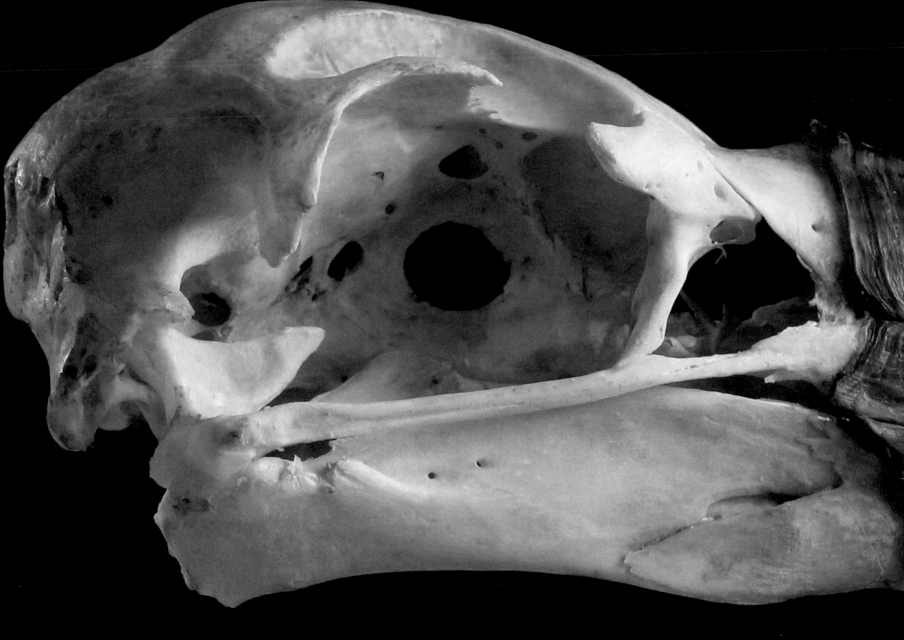

Black-footed Albatross

Bastria nigripes

ALBATROSSES ROAM over vast oceanic ranges.
To locate their prey, they have (unusually for
birds) an acute sense of smell. The long tubules
in the upper mandible play a role in salt secre-

Kingdom: Animalia
Phylum: Chordata
Class: Aves
Order: Ciconiiformes

Family: Procellariidae
Genus: Phoebastria
Behavior: Piscivore/
Diurnal

Wandering Albatross
Diomedea exulans

THE WANDERING ALBATROSS spends almost all of its time at sea, returning to land only to breed. It feeds by using its wide and hooked beak to pluck squid and fish from the surface of the sea at night. More recently, these birds have begun to rely on scavenging the bycatch thrown from trawlers. They drink seawater and must subsequently rid themselves of the excess salt: the tubular nostrils on either side of the bill form part of the excretory apparatus.

Kingdom: Animalia
Phylum: Chordata
Class: Aves
Order: Ciconiiformes
Family: Procellariidae
Genus: Diomedea
Behavior: Piscivore/ Diurnal

DUDLEY'S NOTES:
The wandering albatross may be the biggest flying bird in the world. I have four albatrosses in my collection and this one is, by far, much bigger than the rest. It has a massive beak. It was given to me in this condition and I believe it's a very old antique.

Atlantic Puffin
Fratercula arctica

THIS SKULL RETAINS a shadow of the magnificent red and blue markings that these birds display during the summer months when mates are chosen. Sadly, the color pigments within the beak sheath of bird skulls usually fade over time. The bill is composed of horny plates and the anterior bone is much enlarged and flattened to hold them.

Kingdom: Animalia
Phylum: Chordata
Class: Aves
Order: Ciconiiformes
Family: Laridae
Genus: Fratercula
Behavior: Piscivore/Diurnal

DUDLEY'S NOTES:
Everybody loves puffins, the clowns of the sea, for all the color in their beaks. Unfortunately, the beak doesn't retain its color once it's prepared. It's also very delicate.

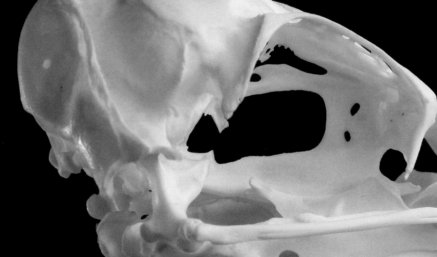

Razorbill
Alca torda

THIS IS A DIVING BIRD: it sits on the surface of the sea, repeatedly dipping its head underwater until it spots a fish (typically a sand eel), which it will chase, catch, and consume before surfacing again. Its rectangular bill is perfect for carrying several sand eels crosswise to feed to its young.

Kingdom: Animalia
Phylum: Chordata
Class: Aves
Order: Ciconiiformes
Family: Laridae
Genus: Alca
Behavior: Piscivore/Diurnal

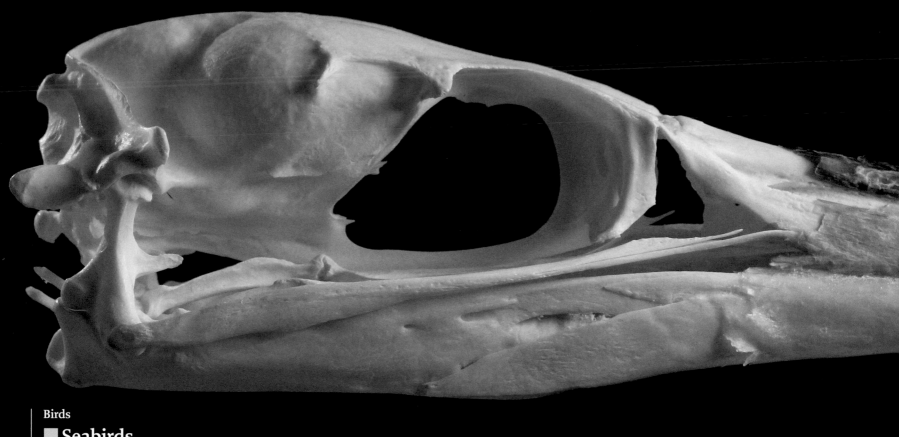

Northern Gannet
Morus bassanus

THE NORTHERN GANNET plunges into the sea from a considerable height like a harpoon to surprise its fish prey, which it consumes before surfacing. Its skull and beak are streamlined to minimize any trauma on impact, and to reduce the splash on entry into the water, giving prey as little warning as possible. The gannet's spiky bill is also important in courtship and display.

Kingdom: Animalia
Phylum: Chordata
Class: Aves
Order: Ciconiiformes

Family: Sulidae
Genus: Morus
Behavior: Piscivore/
Diurnal

Great Cormorant
Phalacrocorax carbo

THIS DIVING BIRD FEEDS on slippery, bottom-dwelling creatures such as flat fish and eels, so its beak has a sharp hook on the end. This allows the bird to better grasp its prey before it can wriggle away. Cormorants can swallow unfeasibly large fish, making them most unpopular amongst anglers.

AKA: Black Shag
Kingdom: Animalia
Phylum: Chordata
Class: Aves
Order: Ciconiiformes
Family: Phalacrocoracidae
Genus: Phalacrocorax
Behavior: Piscivore/Diurnal

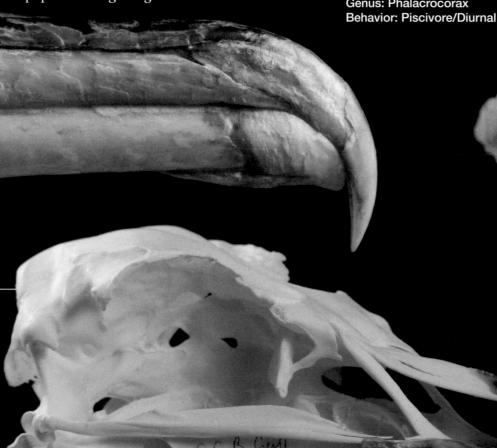

Great Black-backed Gull
Larus marinus

THIS IS THE LARGEST of the gulls, with a robust, strong bill. It indulges in brutish behavior: bullying smaller gulls, gobbling carrion, raiding tips, snatching baby seabirds, and generally living an opportunistic lifestyle. It is found around the coasts of North America, and Europe.

Kingdom: Animalia
Phylum: Chordata
Class: Aves
Order: Ciconiiformes
Family: Laridae
Genus: Larus
Behavior: Carnivore/Diurnal

Southern Giant Petrel
Macronectes giganteus △

THESE BIRDS ARE, as their name suggests, the largest of the petrels. The southern giant petrel skull could be mistaken for that of an albatross. This bird feeds on small fish, krill, and squid, much as do other petrels. Giant petrels are also known to be scavengers and to use their large, heavy, and dangerously hooked beaks to eat other birds.

Kingdom: Animalia
Phylum: Chordata
Class: Aves
Order: Ciconiiformes
Family: Procellariidae
Genus: Macronectes
Behavior: Carnivore/
Diurnal

White-chinned Petrel
Procellaria aequinoctialis ▷

THESE SEABIRDS make regular trips of up to 2,000 kilometers (1,200 miles) in order to feed along nutrient-rich continental shelves. Like their close kin the albatrosses, white-chinned petrels drink seawater; thus, they have similar tubular nostrils to facilitate the excretion of excess salt. The hooked ends of their bills aid them in gripping the wriggling fish that make up much of their diet (they also feed on krill, and some squid).

Kingdom: Animalia
Phylum: Chordata
Class: Aves
Order: Ciconiiformes
Family: Procellariidae
Genus: Procellaria
Behavior: Piscivore/Diurnal

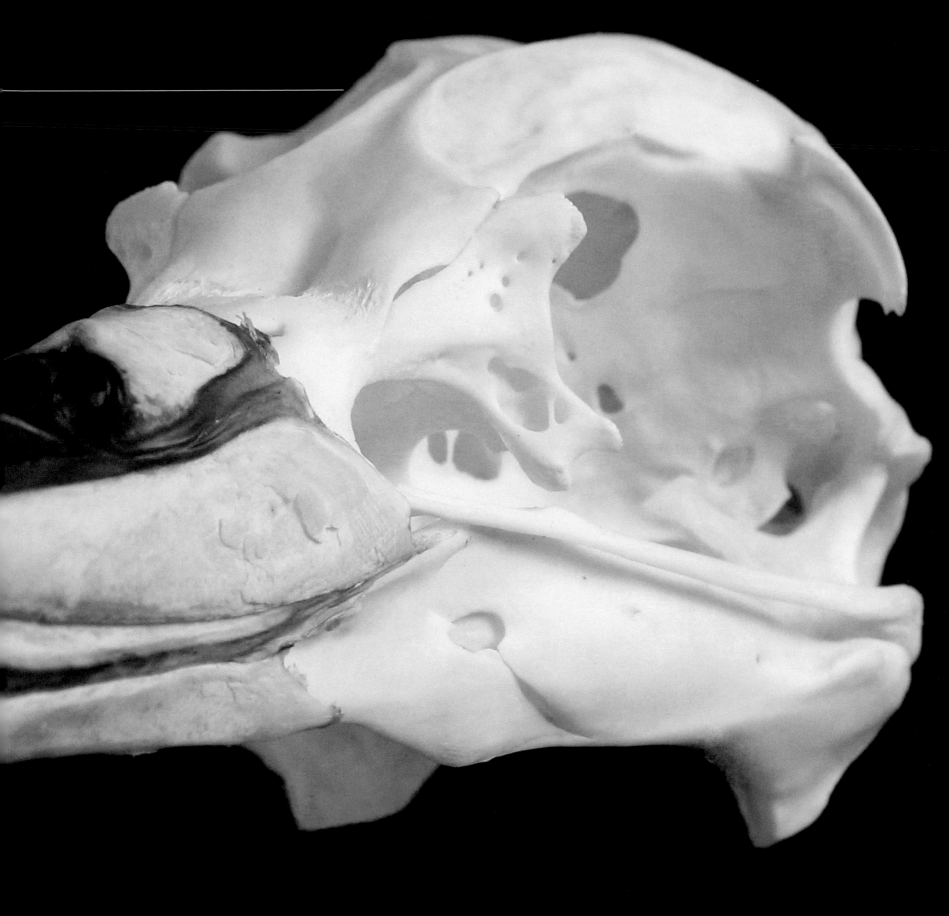

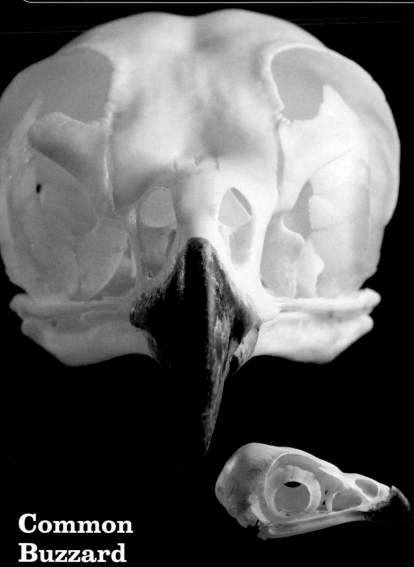

Common Buzzard
Buteo buteo △

THE COMMON BUZZARD is a widespread European diurnal bird of prey. Despite its rather ferocious appearance, it is not the bill that kills the bird's prey but instead the bird's powerful talons; the hooked bill is used to tear off strips of flesh to eat. At the base of the bill is a waxy yellow area known as the cere. Studies have shown that, in many bird species, subtle color changes in the *cere* are used to signal physical fitness and health to potential mates.

Kingdom: Animalia
Phylum: Chordata
Class: Aves
Order: Ciconiiformes
Family: Accipitridae
Genus: Buteo
Behavior: Carnivore/Diurnal

Lappet-faced Vulture
Torgos tracheliotus △

THE LAPPET-FACED VULTURE is large enough to defend carcasses from terrestrial scavengers such as jackals. The squarish and thick-set skull bears a powerful beak that can rip through the tough hide of dead animals, allowing smaller vultures to gain access to the meat within. In times of need it can also employ its fearsome beak to kill weak or injured animals.

Kingdom: Animalia
Phylum: Chordata
Class: Aves
Order: Ciconiiformes
Family: Accipitridae
Genus: Torgos
Behavior: Carnivore/Diurnal

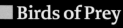

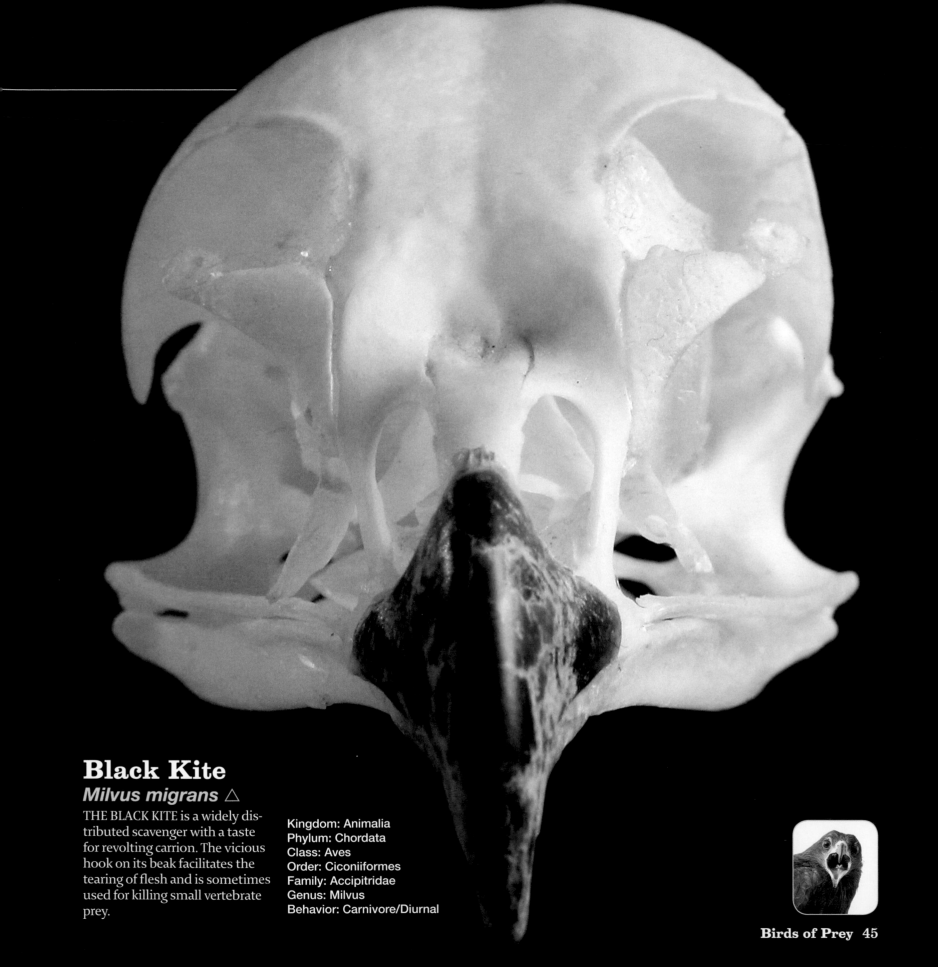

Black Kite

Milvus migrans △

THE BLACK KITE is a widely distributed scavenger with a taste for revolting carrion. The vicious hook on its beak facilitates the tearing of flesh and is sometimes used for killing small vertebrate prey.

Kingdom: Animalia
Phylum: Chordata
Class: Aves
Order: Ciconiiformes
Family: Accipitridae
Genus: Milvus
Behavior: Carnivore/Diurnal

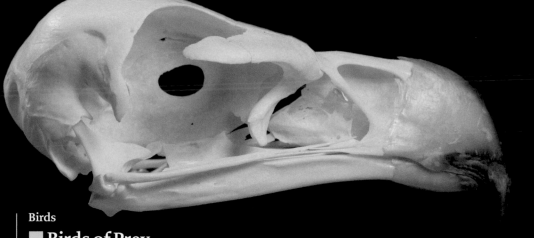

White-tailed Eagle
△ *Haliaeetus albicilla*

THE WHITE-TAILED EAGLE will opportunistically scavenge year-round; however, the enormous hooked beak, and wicked talons are also put to use killing birds, fish, and small mammals. The enormous orbits house the eagle's large eyes. The forward-facing eyes give the eagle binocular vision and depth perception, essential for scooping fish from the surface of the sea.

Kingdom: Animalia
Phylum: Chordata
Class: Aves
Order: Ciconiiformes
Family: Accipitridae
Genus: Haliaeetus
Behavior: Carnivore/Diurnal

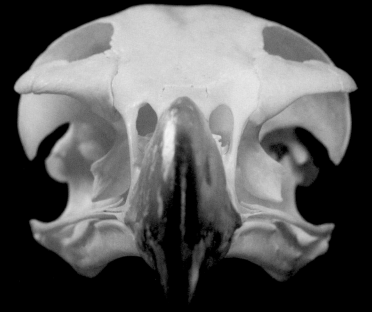

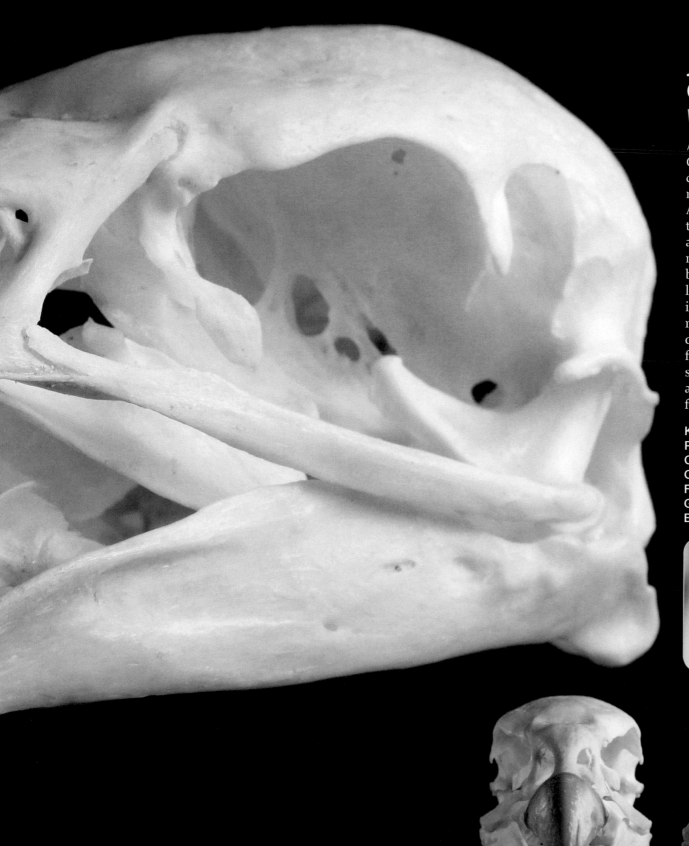

Andean Condor
Vultur gryphus ◁

ALTHOUGH NOT AS RARE as the Californian condor, the andean condor is endangered. It is the national symbol of six South American countries and has the second-largest wingspan of any bird (up to ten feet or three meters—only the wandering albatross is larger) and one of the longest lifespans (over fifty years in the wild). As its scientific name suggests, the andean condor is a member of the vulture family. Like other vultures, it is a scavenger, lacks head plumage, and has a sharply pointed beak for tearing into flesh.

Kingdom: Animalia
Phylum: Chordata
Class: Aves
Order: Ciconiiformes
Family: Ciconiidae
Genus: Vultur
Behavior: Carnivore/Diurnal

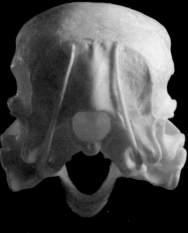

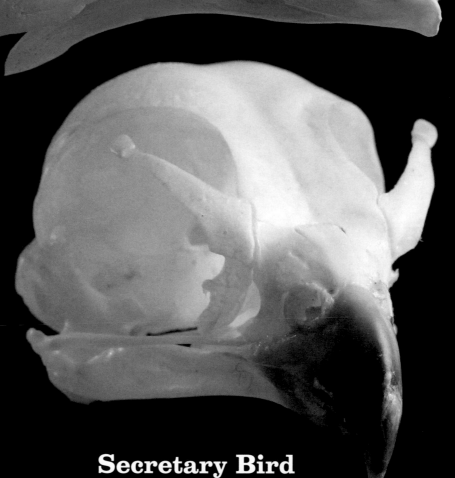

Turkey Vulture
▷ *Cathartes aura*

THE UPPER MANDIBLE of the turkey vulture is wickedly hooked to tear into carrion. The region that contains the olfactory lobe of the brain is particularly large: this bird has a well-developed sense of smell (an uncommon trait in birds) that it uses to locate dead animals.

Kingdom: Animalia
Phylum: Chordata
Class: Aves
Order: Ciconiiformes
Family: Ciconiidae
Genus: Cathartes
Behavior: Carnivore/ Diurnal

Common Kestrel
▷ *Falco tinnunculus*

THE KESTREL SKULL, with a sharply hooked beak and large volume given over to housing the eyes, is that of a typical raptor. For an aerial predator, the common kestrel hunts at a relatively low altitude. It skilfully hovers at about twenty meters (sixty-five feet) above the ground before diving down to catch small mammals, lizards or song birds.

Kingdom: Animalia
Phylum: Chordata
Class: Aves
Order: Ciconiiformes
Family: Falconidae
Genus: Falco
Behavior: Carnivore/Diurnal

Secretary Bird
◁ *Sagittarius serpentarius*

THIS AFRICAN BIRD OF PREY has an eaglelike body that sits atop crane-like legs. It is a largely terrestrial hunter, stalking its prey on foot. The skull closely resembles that of any eagle, with a powerful, hooked beak. It has a reputation for feeding on snakes but will also take small mammals, and insects.

Kingdom: Animalia
Phylum: Chordata
Class: Aves
Order: Ciconiiformes
Family: Sagittariidae
Genus: Sagittarius
Behavior: Carnivore/ Diurnal

Helmeted Guineafowl
Numida meleagris ◁

THE UNMISTAKEABLE DUMPY, small-headed profile of the helmeted guineafowl is familiar in many countries; these birds are commonly domesticated. They sport a fetching casque, the purpose of which is unclear.

Kingdom: Animalia
Phylum: Chordata
Class: Aves
Order: Galliformes
Family: Numidiae
Genus: Numida
Behavior: Omnivore/Diurnal

Himalayan Monal Pheasant
Lophophorus impejanus △

THE MALE OF THIS species is a spectacular metallic green, blue, and red with a long crest on its head, and was the inspiration for the two meters (seventy-six inches) tall, flightless bird in the animated film *Up*. However, this bird is a more modest seventy centimeters (twenty-eight inches) from beak to tail. The robust beak with its broad tip is used to dig for tubers, and worms beneath the soil in frequently frozen environments.

Kingdom: Animalia
Phylum: Chordata
Class: Aves
Order: Galliformes
Family: Phasianidae

Genus: Lophophorus
Behavior: Omnivore/ Diurnal

Red-legged Seriema
Cariama cristata △

THE RED-LEGGED SERIEMA is a predominantly terrestrial bird, found in South America. It has a sharp hooked beak, reminiscent of a raptor. This bird has a curious way of devouring prey, such as frogs and lizards, that are too big to be swallowed whole: it grasps them firmly in its beak and uses a specialized sickle-shaped claw to tear them into pieces.

Kingdom: Animalia
Phylum: Chordata
Class: Aves
Order: Gruiformes

Family: Cariamidae
Genus: Cariama
Behavior: Omnivore/ Diurnal

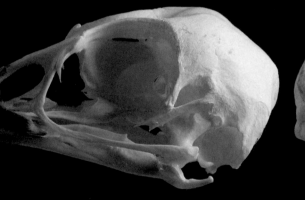

Capercaillie
Tetrao urogallus

THE CAPERCAILLIE is a turkey-sized bird of European forests. It has a diet of pine needles and bilberries. When still clothed in flesh and feathers, the males have a striking patch of red skin above the eye, and perform an extraordinary mating display.

Kingdom: Animalia

Phylum: Chordata

Class: Aves

Order: Galliformes

Family: Phasianidae

Genus: Tetrao

Behavior: Herbivore/

Diurnal

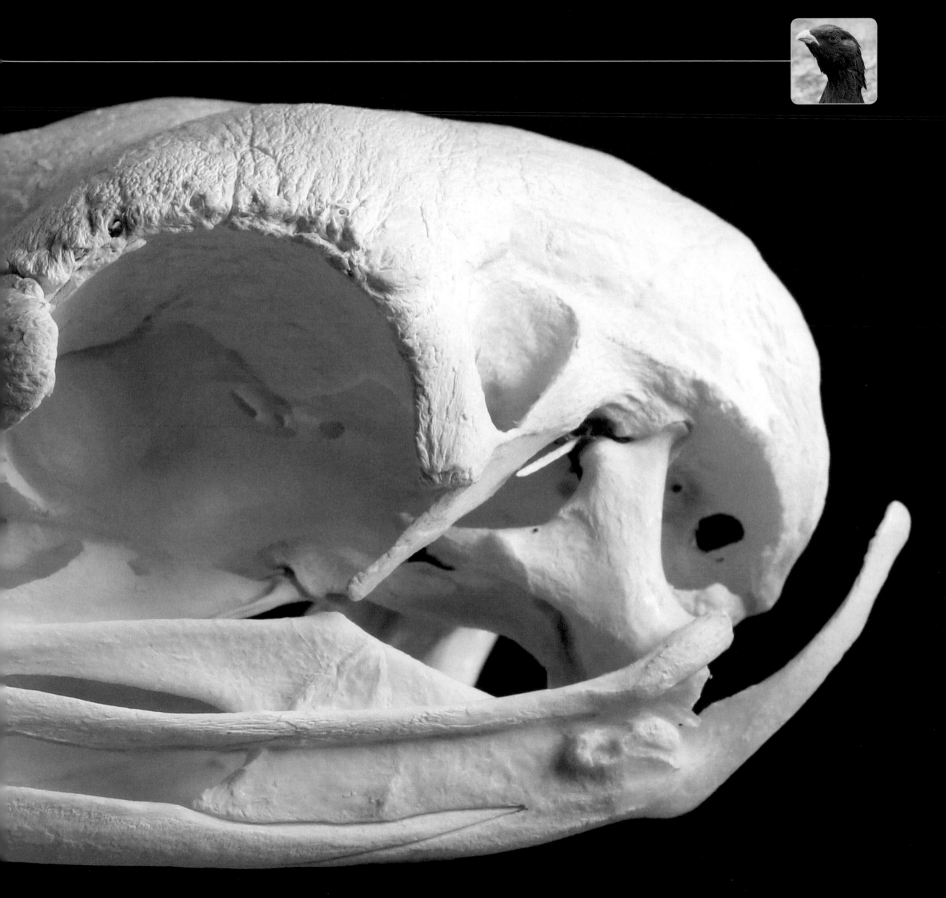

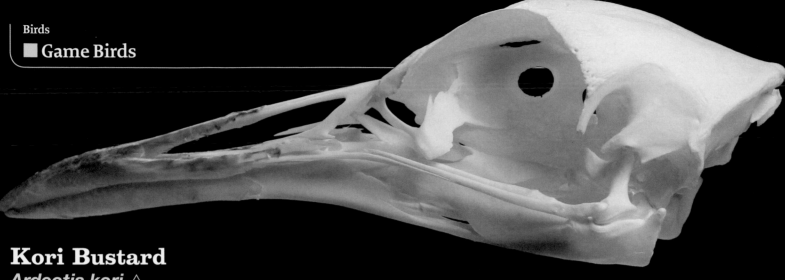

Kori Bustard
Ardeotis kori △

THIS BIRD STRIDES through the short grass of open savannah regions, its height affording it a good view of approaching predators while looking for insects and small reptiles or mammals to eat. The kori bustard is the heaviest flying bird.

Kingdom: Animalia
Phylum: Chordata
Class: Aves
Order: Gruiformes
Family: Otididae
Genus: Ardeotis
Behavior: Carnivore/Diurnal

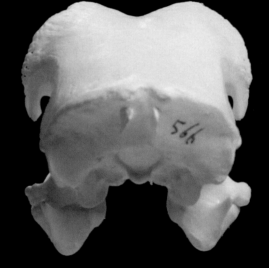

Razor-billed Curassow
Mitu tuberosa ▽

THE RAZOR-BILLED CURASSOW is named for the swollen base of its upper mandible. Living a lifestyle similar to that of the helmeted curassow, it feeds on fallen fruit or in the trees on fresh fruit and buds. A plump and tasty species, curassows have long been hunted by indigenous peoples. Although human population growth and habitat destruction have taken their toll, this species is still thriving.

Kingdom: Animalia
Phylum: Chordata
Class: Aves
Order: Craciformes
Family: Cracidae
Genus: Mitu
Behavior: Omnivore/Diurnal

Helmeted Curassow
Pauxi pauxi

THESE BIRDS FORAGE for fallen fruit and seeds on the forest floor in the mountainous cloud forests of South America. The ornamental fig-shaped casque, colored grayish blue in life, is spectacular. The purpose of the casque is unclear, but the male helmeted curassow produces a low, booming call, which suggests that the casque has a resonating function.

Kingdom: Animalia
Phylum: Chordata
Class: Aves
Order: Craciformes
Family: Cracidae
Genus: Pauxi
Behavior: Omnivore/ Diurnal

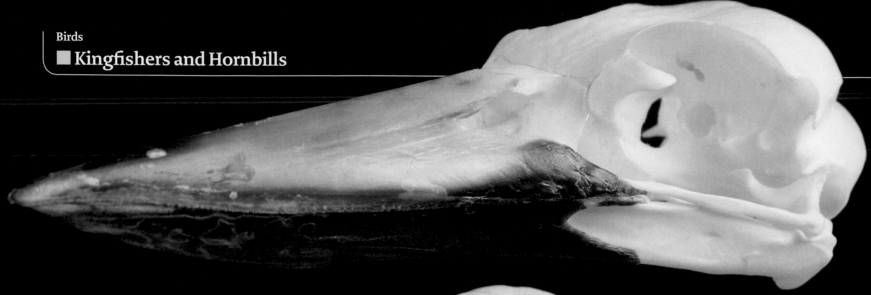

Blue-breasted Kingfisher
Halcyon malimbica ▷

THE BLUE-BREASTED kingfisher has a long, daggerlike bill to seize insects and frogs. It is one of the wood kingfishers and doesn't routinely catch fish, but instead drops from a perch to the ground, ambushing its prey.

Kingdom: Animalia
Phylum: Chordata
Class: Aves
Order: Coraciiformes
Family: Alcedinidae
Genus: Halcyon
Behavior: Carnivore/ Diurnal

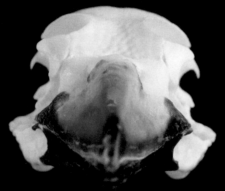

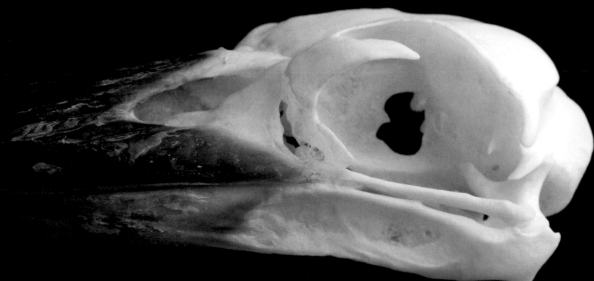

Laughing Kookaburra
Dacelo novaeguineae △

THE KOOKABURRAS are a group of kingfishers native to Australia that have very little dependence on water. The laughing kookaburra hunts from a perch, as do other kingfishers, but swoops down on small mammals, lizards, and even venomous snakes. The beak shape is characteristically that of a kingfisher.

Kingdom: Animalia
Phylum: Chordata
Class: Aves
Order: Coraciiformes
Family: Alcedinidae

Genus: Dacelo
Behavior: Carnivore/ Diurnal

Black-casqued Hornbill
Ceratogymna atrata

THE CASQUE OF THIS MALE spans much of the length of the beak (it is much reduced but still apparent in females). The hollow space of the casque is connected to the bird's mouth, and therefore it is likely that the casque acts to attract females visually, and as a resonator of vocalizations. The characteristic downturned beak is employed in the consumption of a broad diet: from seeds and fruit to insects and small mammals.

Kingdom: Animalia
Phylum: Chordata
Class: Aves
Order: Bucerotiformes
Family: Bucerotidae
Genus: Ceratogymna
Behavior: Omnivore/
Diurnal

Abyssinian Ground Hornbill
Bucorvus abyssinicus ▷

BECAUSE OF THE extraordinary variety of their beaks, Dudley has an inordinate fondness for hornbills; there are more than twenty-five in his collection. All are characterized by an elaborate, keratinous casque on their upper mandible. In the Abyssinian ground hornbill, the casque takes the form of an open-ended quadrant. Usually filled with a spongy, porous matrix, it is thought that casques act as a resonating chamber. They may also play a role in species recognition and in signaling sexual maturity.

Kingdom: Animalia
Phylum: Chordata
Class: Aves
Order: Bucerotiformes
Family: Bucorvidae
Genus: Bucorvus
Behavior:
Omnivore/Diurnal

African Pied Hornbill
Tockus fasciatus ▽

THE BILL OF THE African pied hornbill is similar to that of the toucans in that it has a serrated margin to its mandibles to afford a better grip on its fruit diet (although it is also partial to insects). In this species the casque reaches almost to the tip of the bill. Male casques are larger than those of the female, and in sexually immature individuals the casque is not present.

Kingdom: Animalia
Phylum: Chordata
Class: Aves
Order: Bucerotiformes
Family: Bucerotidae
Genus: Tockus
Behavior: Omnivore/
Diurnal

Brown-cheeked Hornbill
Bycanistes cylindricus ▷

THIS HORNBILL has an enormous casque, larger than the beak itself. The large casque is thought to have a resonating function, helping to produce the very loud calls characteristic of this genus. This species has near threatened status on the International Union for Conservation of Nature (IUCN) Red List, as its numbers are in rapid decline due to habitat destruction.

Kingdom: Animalia
Phylum: Chordata
Class: Aves
Order: Bucerotiformes
Family: Bucerotidae
Genus: Bycanistes
Behavior: Omnivore/ Diurnal

Blyth's Hornbill
Rhyticeros plicatus ▷

THE LOW "WREATH" casque that rests on the upper mandible of the Blyth's hornbill is typical of its genus. The downwardly curved bill is a classic shape, common to all hornbills and suited to this one's diet of fruit (mainly figs), beefed up with occasional snacks of small animals and insects.

Kingdom: Animalia
Phylum: Chordata
Class: Aves
Order: Bucerotiformes
Family: Bucerotidae
Genus: Rhyticeros
Behavior: Omnivore/Diurnal

Eastern Yellow-billed Hornbill
Tockus flavirostris ◁

THE EASTERN YELLOW-BILLED hornbill has a wonderful foraging relationship with mongooses. In this mutually beneficial association, the hornbills feed on the insects that the mongooses disturb, while the mongooses benefit from the bird's vigilance for predators. The mongoose will wait for the hornbill before beginning foraging, but if hornbills are ready first, they will call into the mongoose burrows to hurry them along.

Kingdom: Animalia
Phylum: Chordata
Class: Aves
Order: Bucerotiformes
Family: Bucerotidae
Genus: Tockus
Behavior: Omnivore/ Diurnal

Great Hornbill
Buceros bicornis

HERE WE HAVE a skull complete with skeleton, mounted on a perch that presented some serious balance problems for our photographer—and, presumably, for the bird as well. Many large-billed birds require extra musculature to support their heads. This great hornbill is found throughout Southeast Asia.

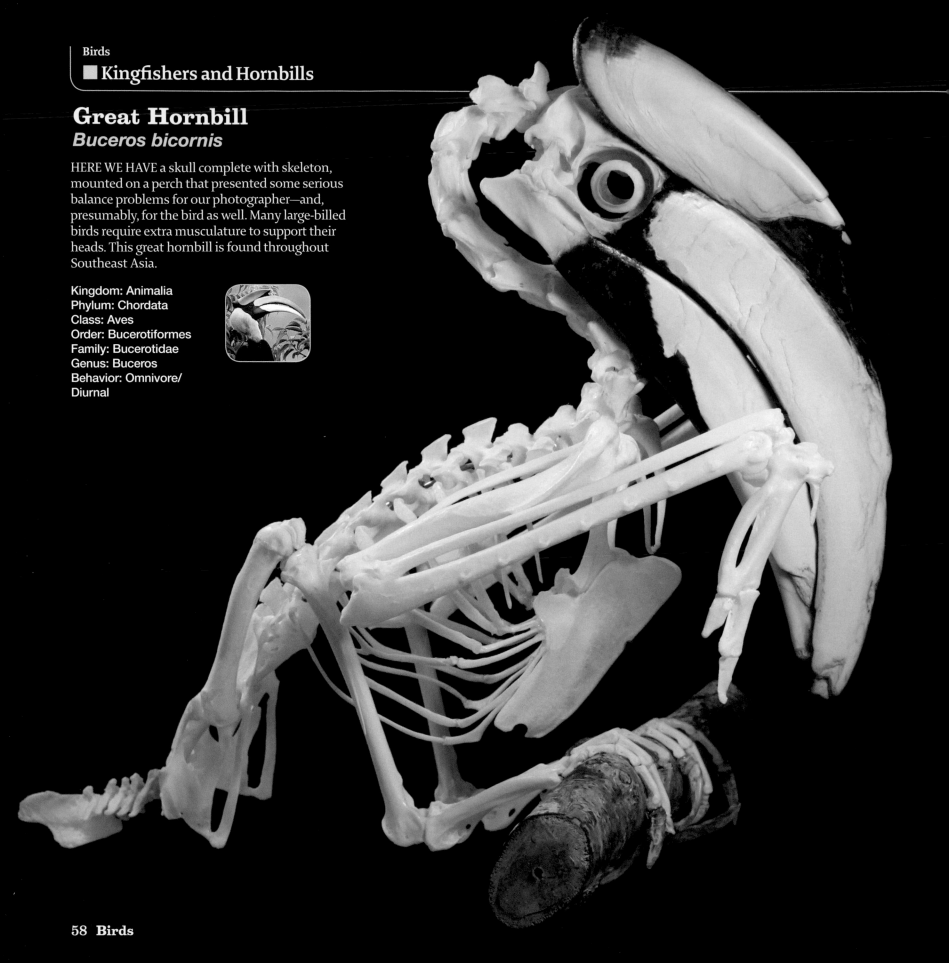

Kingdom: Animalia
Phylum: Chordata
Class: Aves
Order: Bucerotiformes
Family: Bucerotidae
Genus: Buceros
Behavior: Omnivore/ Diurnal

Red-billed Hornbill
Tockus erythrorhynchus

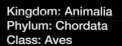

THIS IS A SMALL SPECIES of hornbill—about fifty centimeters (twenty inches) from beak to tip of tail—found in sub-Saharan Africa. It has a much reduced casque, just a slight protrusion on the upper mandible. It occasionally feeds on fruit, but more often on insects, using its long down-curved bill to dig for them in the dirt.

Kingdom: Animalia
Phylum: Chordata
Class: Aves
Order: Bucerotiformes
Family: Bucerotidae
Genus: Tockus
Behavior: Omnivore/Diurnal

Helmeted Hornbill
Rhinoplax vigil

UNIQUELY, THIS BIRD'S casque is not a lightweight, mostly hollow space but rather a massive ivory block constituting around 10 percent of the its body weight. The helmeted hornbill's casque is stained red by waxy secretions from the bird's preen gland. The heavy casque is used in aerial jousts between males contesting resources. The photograph shows a juvenile bird.

Kingdom: Animalia
Phylum: Chordata
Class: Aves
Order: Bucerotiformes
Family: Bucerotidae
Genus: Rhinoplax
Behavior: Omnivore/Diurnal

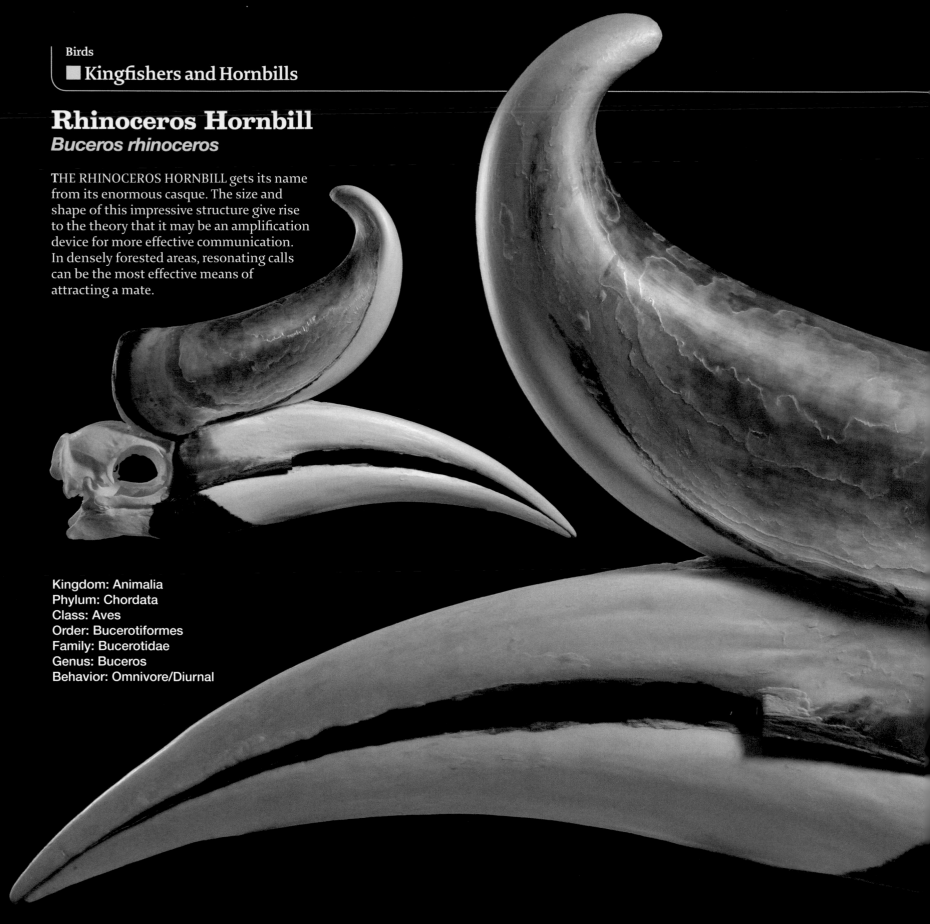

Rhinoceros Hornbill
Buceros rhinoceros

THE RHINOCEROS HORNBILL gets its name from its enormous casque. The size and shape of this impressive structure give rise to the theory that it may be an amplification device for more effective communication. In densely forested areas, resonating calls can be the most effective means of attracting a mate.

Kingdom: Animalia
Phylum: Chordata
Class: Aves
Order: Bucerotiformes
Family: Bucerotidae
Genus: Buceros
Behavior: Omnivore/Diurnal

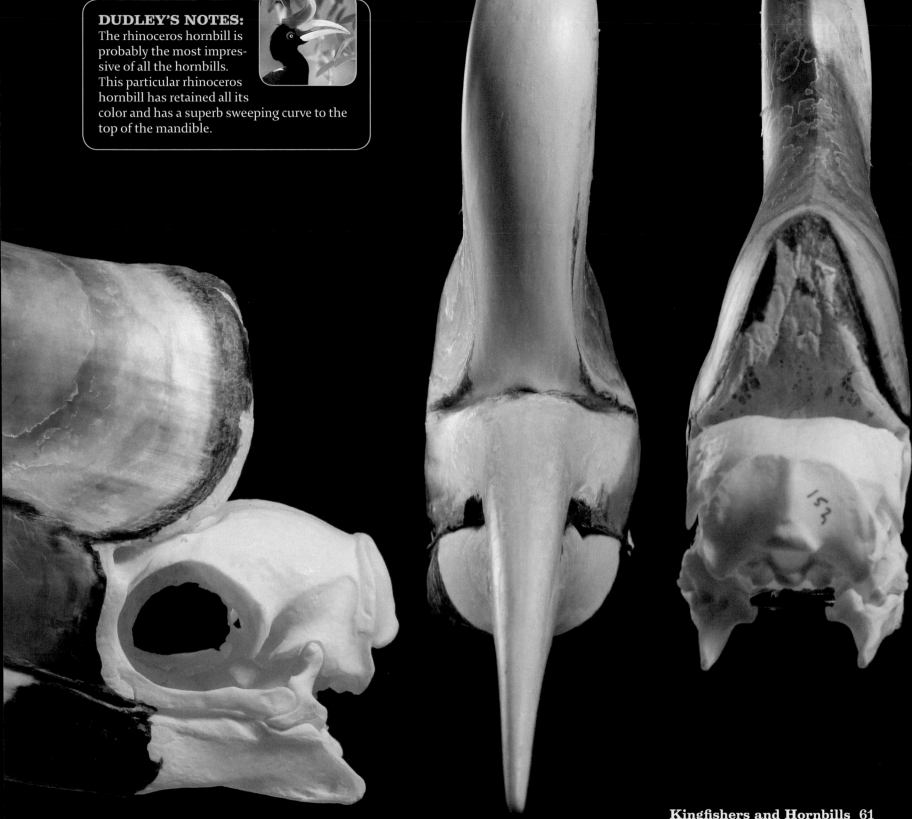

Rufous Hornbill
▷ *Buceros hydrocorax*

WHEN MATURE, THE BEAK, and casque of the rufous hornbill are a bright red. The bill is serrated, in common with those of other frugivorous species, to allow better manipulation of the fruit that are its primary diet. It is endemic to the Philippines; there are several subspecies found on different islands of the Philippine archipelago. The image here is of a juvenile bird.

AKA: Philippine Hornbill
Kingdom: Animalia
Phylum: Chordata
Class: Aves

Order: Bucerotiformes
Family: Bucerotidae
Genus: Buceros
Behavior: Omnivore/ Diurnal

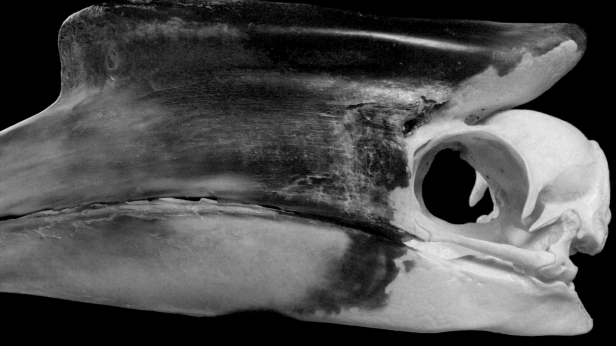

Tarictic Hornbill
▷ *Penelopides sp.*

AS IS TRUE of many other predominantly frugivorous birds, this hornbill's mandibles have serrated margins. The casque of this species is a simple keratinous protrusion. This genus is subject to much debate amongst taxonomists, and it is unclear exactly which species this skull belonged to.

Kingdom: Animalia
Phylum: Chordata
Class: Aves
Order: Bucerotiformes
Family: Bucerotidae
Genus: Penelopides
Behavior: Omnivore/ Diurnal

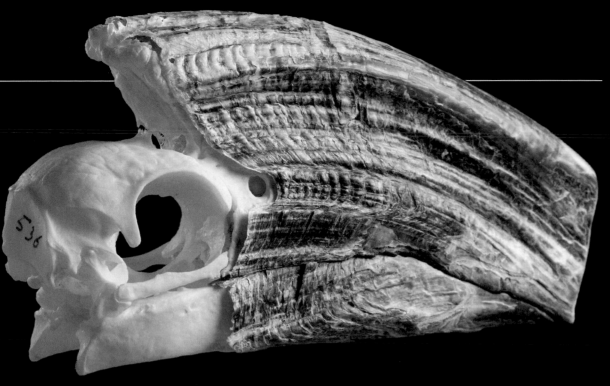

Trumpeter Hornbill
◁ *Bycanistes bucinator*

THE TRUMPETER hornbill sports a massive casque—larger in the males than it is in the females, so perhaps an indicator of sexual fitness or social status. As suggested in similar hornbills, the large casque may have a resonating function for vocalization. The trumpeter hornbill is widespread in Africa.

Kingdom: Animalia
Phylum: Chordata
Class: Aves
Order: Bucerotiformes
Family: Bucerotidae
Genus: Rhyticeros
Behavior: Omnivore/Diurnal

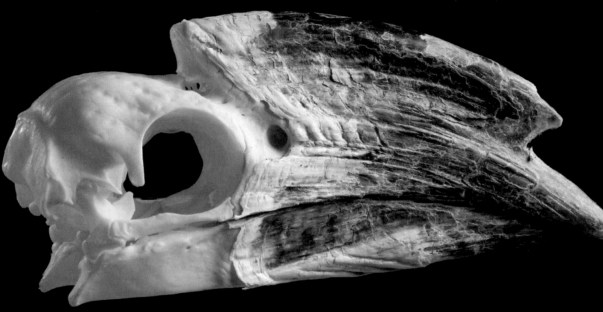

Silvery-cheeked Hornbill
△ *Bycanistes brevis*

THIS EAST AFRICAN hornbill has a large, protruding casque widely thought to be a vocal amplifier of sorts. Females have a smaller casque than males (this is a female's skull), and the casque may also play a role in sexual selection.

Kingdom: Animalia
Phylum: Chordata
Class: Aves
Order: Bucerotiformes

Family: Bucerotidae
Genus: Bicanistes
Behavior: Omnivore/ Diurnal

Southern Ground Hornbill
Bucorvus leadbeateri ▷

THIS IS THE LARGEST of the African hornbill species, reaching weights of over six kilograms (thirteen pounds). Its casque is diminished and serves primarily to strengthen the upper mandible. The gap between the upper and lower mandibles means that the force of its bite is transferred to the tip. The bird can use this tweezerlike action to keep dangerous ▨▨ (such as poisonous snakes) ▨▨ a safe distance while squeezing them to death.

Kingdom: Animalia
Phylum: Chordata
Class: Aves
Order: Bucerotiformes
Family: Bucorvidae
Genus: Bucorvus
Behavior: Omnivore/Diurnal

Wreathed Hornbill
Rhyticeros undulatus ▷

THE WRINKLED CASQUE on the upper mandible of this bird is thought to be a species recognition cue. This skull is likely to be that of a relatively young bird: the wrinkles on its upper mandible (this individual had only one) are rough signifiers of age. Wreathed hornbills are frugivorous birds of Southeast Asia.

Kingdom: Animalia
Phylum: Chordata
Class: Aves
Order: Bucerotiformes
Family: Bucerotidae
Genus: Rhyticeros
Behavior: Omnivore/
Diurnal

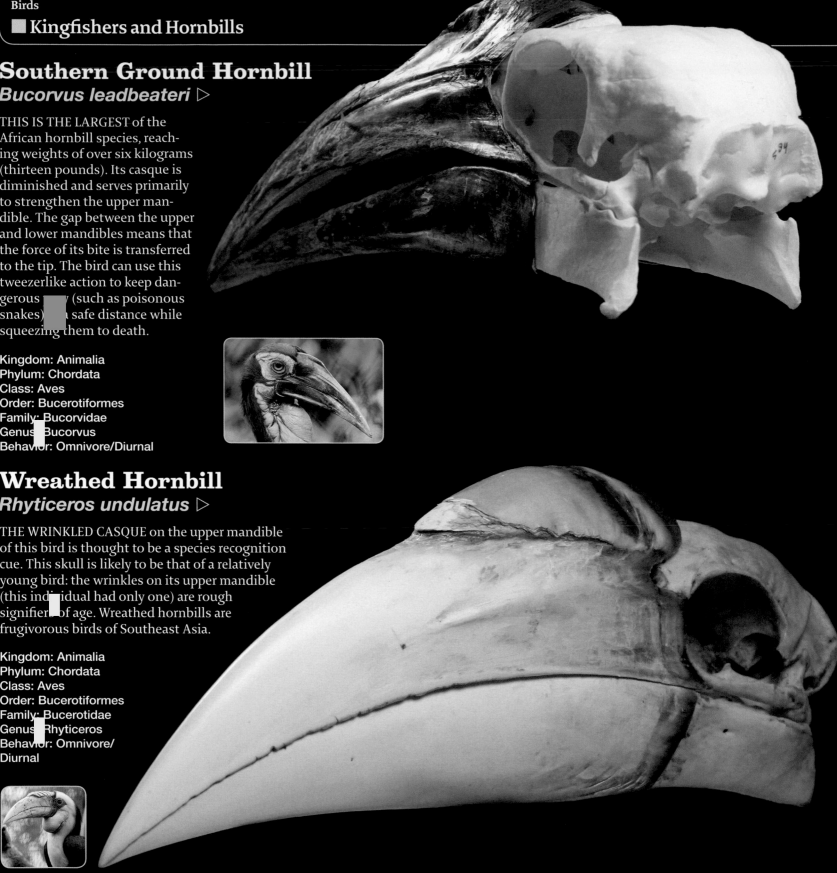

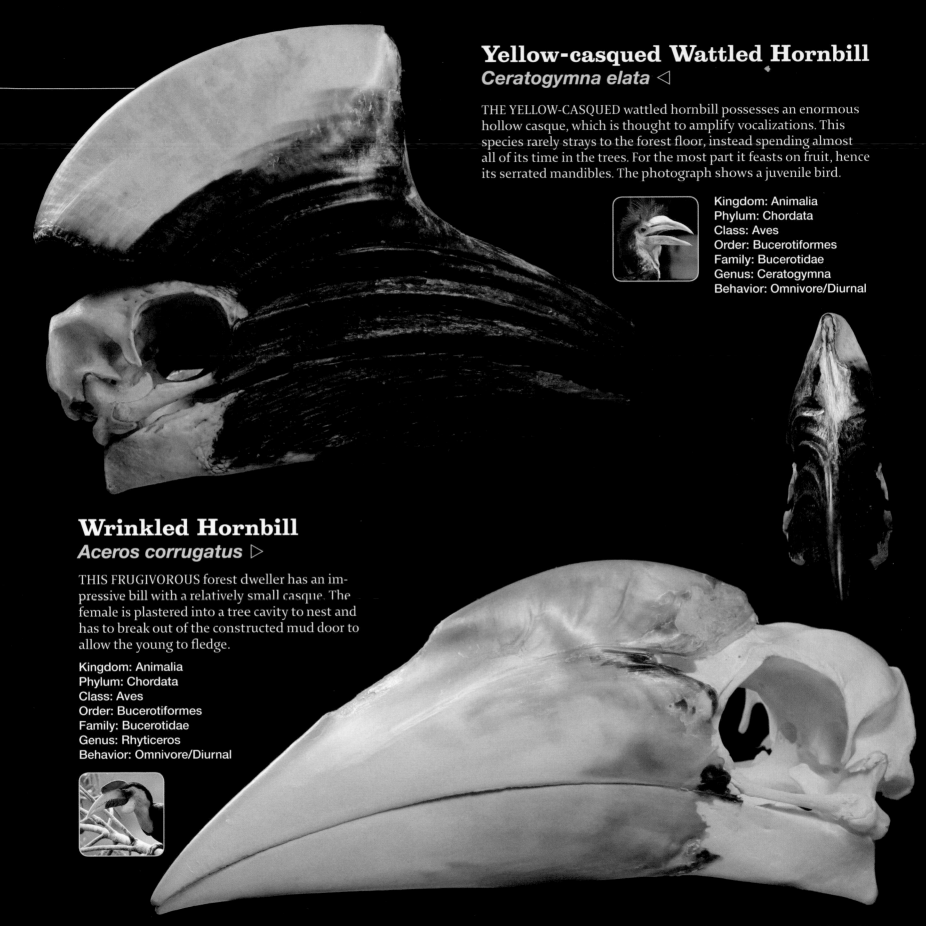

Yellow-casqued Wattled Hornbill
Ceratogymna elata ◁

THE YELLOW-CASQUED wattled hornbill possesses an enormous hollow casque, which is thought to amplify vocalizations. This species rarely strays to the forest floor, instead spending almost all of its time in the trees. For the most part it feasts on fruit, hence its serrated mandibles. The photograph shows a juvenile bird.

Kingdom: Animalia
Phylum: Chordata
Class: Aves
Order: Bucerotiformes
Family: Bucerotidae
Genus: Ceratogymna
Behavior: Omnivore/Diurnal

Wrinkled Hornbill
Aceros corrugatus ▷

THIS FRUGIVOROUS forest dweller has an impressive bill with a relatively small casque. The female is plastered into a tree cavity to nest and has to break out of the constructed mud door to allow the young to fledge.

Kingdom: Animalia
Phylum: Chordata
Class: Aves
Order: Bucerotiformes
Family: Bucerotidae
Genus: Rhyticeros
Behavior: Omnivore/Diurnal

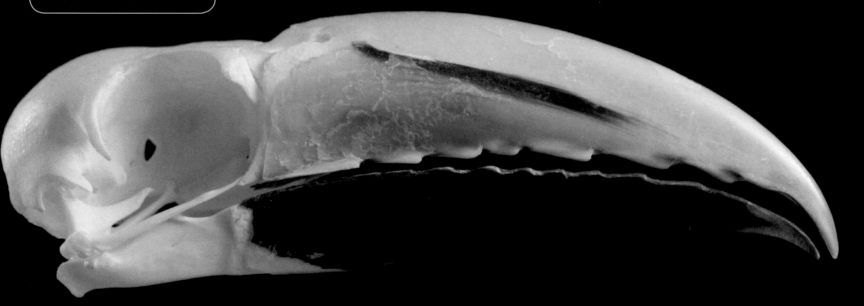

Plate-billed Mountain Toucan
Andigena laminirostris ▷

A DEDICATED FRUGIVORE, the plate-billed mountain toucan has spectacular mandibles with serrated edges to improve its grip on fruit. It has a very limited distribution in the andes; habitat destruction coupled with global climate change are bad news for the population of this species.

Kingdom: Animalia
Phylum: Chordata
Class: Aves
Order: Piciformes
Family: Ramphastidae
Genus: Andigena
Behavior: Frugivore/Diurnal

DUDLEY'S NOTES:
I particularly like this toucan because it has a lot of colors in the beak. Toucan skulls are among my favorite to collect because there are so many different varieties.

Green Aracari
Pteroglossus viridis △

THESE SMALL TOUCANS live in tree hollows (usually ones previously occupied by woodpeckers). The green aracari consumes a wide range of different fruits in the canopy of tropical forests. The serrated margin of its mandibles helps the toucan to grip fruit as it forages.

Kingdom: Animalia
Phylum: Chordata
Class: Aves
Order: Piciformes
Family: Ramphastidae
Genus: Pteroglossus
Behavior: Frugivore/Diurnal

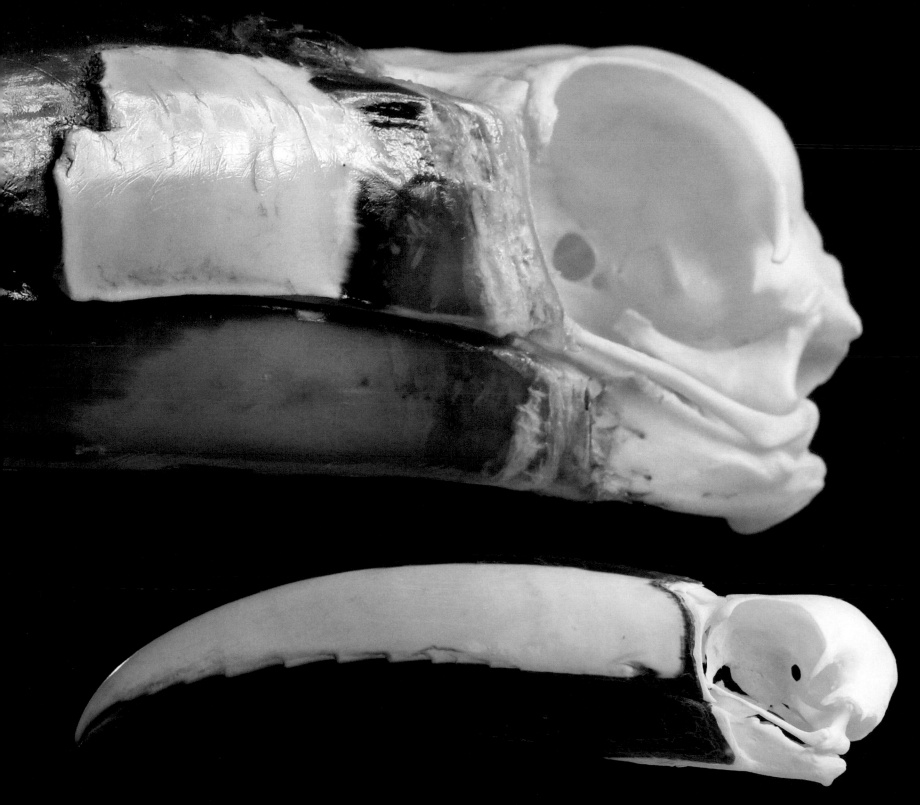

Black-necked Aracari
Pteroglossus aracari △

THE DISPROPORTIONATELY large beak of this toucan (and all others) is likely to have at least some thermoregulatory function. It is well supplied with blood and dissipates heat when the bird is warm. However, it is also a useful adaptation for plucking and manipulating the fruit these birds eat.

Kingdom: Animalia
Phylum: Chordata
Class: Aves
Order Piciformes
Family Ramphastidae
Genus: Pteroglossus
Behavior: Frugivore/Diurnal

■ Toucans and Woodpeckers

Spot-billed Toucanet
Selenidera maculirostris ▷

AS WITH OTHER TOUCANS, you can see the serrated margin of its mandibles that help the bird grip and manipulate fruit. The pattern of black and white on the toucanet's bill (from which it derives its name) is peculiar to every individual and so can be used as a fingerprint for each bird.

Kingdom: Animalia
Phylum: Chordata
Class: Aves
Order: Piciformes
Family: Ramphastidae
Genus: Selenidera
Behavior: Omnivore/Diurnal

Toco Toucan
Ramphastos toco ▽

THE TOCO TOUCAN is the classic toucan species of Guinness fame. It has a black feathered body with a white throat, and an enormous, bright orange beak. As with other toucans the large bill is likely to be involved in thermoregulation and the serrated margin of the mandibles helps to grip the fruits that make up the bird's diet.

Kingdom: Animalia
Phylum: Chordata
Class: Aves
Order: Piciformes

Family: Ramphastidae
Genus: Ramphastos
Behavior: Omnivore/
Diurnal

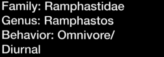

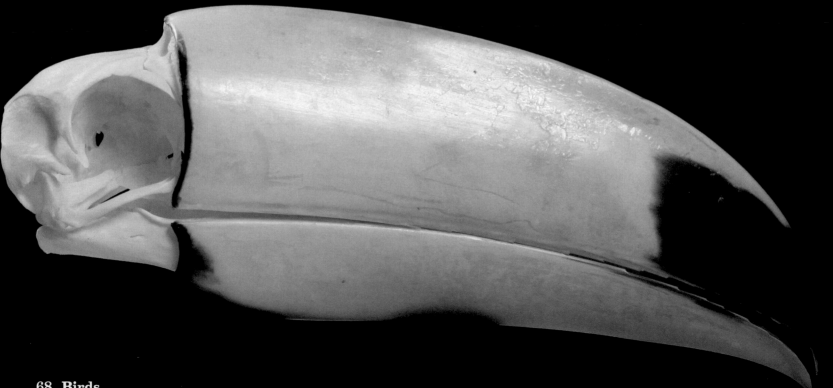

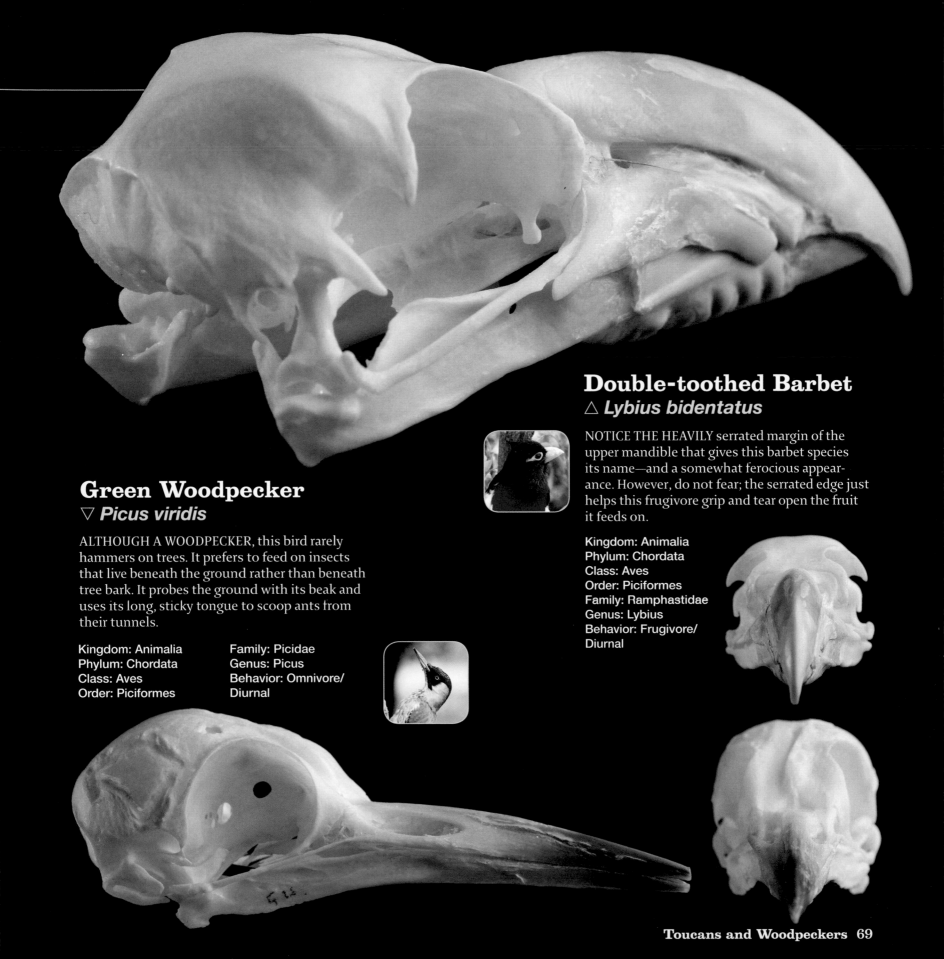

Green Woodpecker
▽ *Picus viridis*

ALTHOUGH A WOODPECKER, this bird rarely hammers on trees. It prefers to feed on insects that live beneath the ground rather than beneath tree bark. It probes the ground with its beak and uses its long, sticky tongue to scoop ants from their tunnels.

Kingdom: Animalia
Phylum: Chordata
Class: Aves
Order: Piciformes

Family: Picidae
Genus: Picus
Behavior: Omnivore/ Diurnal

Double-toothed Barbet
△ *Lybius bidentatus*

NOTICE THE HEAVILY serrated margin of the upper mandible that gives this barbet species its name—and a somewhat ferocious appearance. However, do not fear; the serrated edge just helps this frugivore grip and tear open the fruit it feeds on.

Kingdom: Animalia
Phylum: Chordata
Class: Aves
Order: Piciformes
Family: Ramphastidae
Genus: Lybius
Behavior: Frugivore/ Diurnal

European Nightjar
Caprimulgus europaeu

TYPICALLY, THE NIGHTJAR will perch around an open area, after dark, where it can watch for insects. It will periodically (and acrobatically) fly from its perch and catch a flying moth or beetle. This task is facilitated by a short but extremely wide bill. The living bird has a series of bristles on either side of the mouth that trap any insects that might otherwise slip out.

Kingdom: Animalia
Phylum: Chordata
Class: Aves
Order: Strigiformes
Family: Caprimulgidae
Genus: Caprimulgus
Behavior: Insectivore/Nocturnal

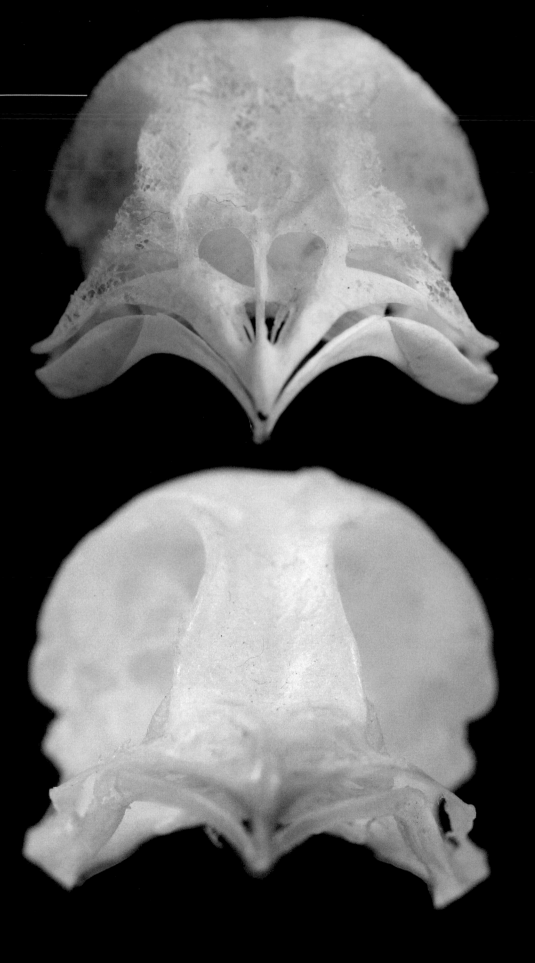

Common Swift
Apus apus

THE COMMON SWIFT has a broad beak and an enormous gape to catch flying invertebrates on the wing. This species spends a very large proportion of its life flying, never perching except to nest. Individual birds feed, mate, drink, and even sleep while flying.

Kingdom: Animalia
Phylum: Chordata
Class: Aves
Order: Apodiformes
Family: Apodidae
Genus: Apus
Behavior: Insectivore/Diurnal

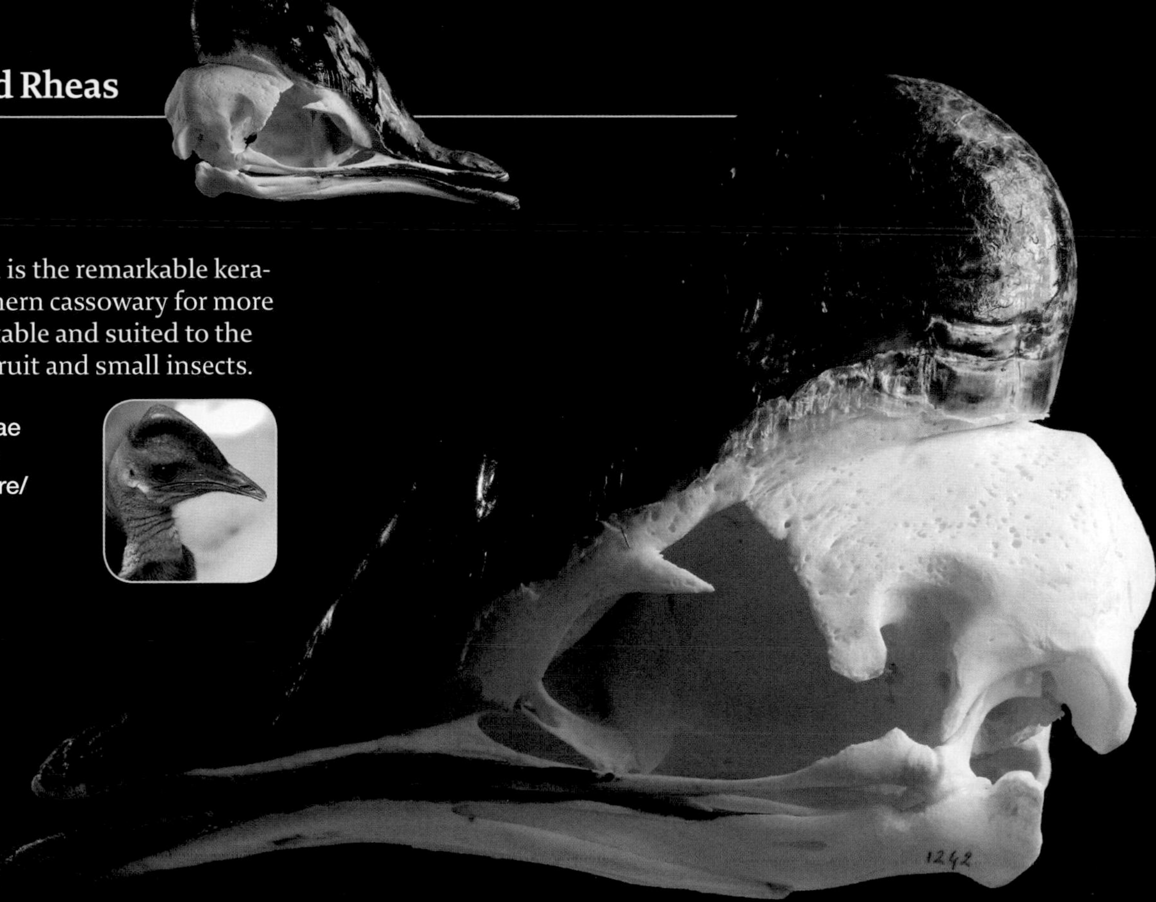

Dwarf Cassowary
Casuarius bennetti ▷

THE FIRST THING one notices on this skull is the remarkable keratinous casque (see the closely related southern cassowary for more details). The beak morphology is unremarkable and suited to the dwarf cassowary's generalist diet of fallen fruit and small insects.

Kingdom: Animalia
Phylum: Chordata
Class: Aves
Order: Struthioniformes

Family: Casuariidae
Genus: Casuarius
Behavior: Frugivore/
Diurnal

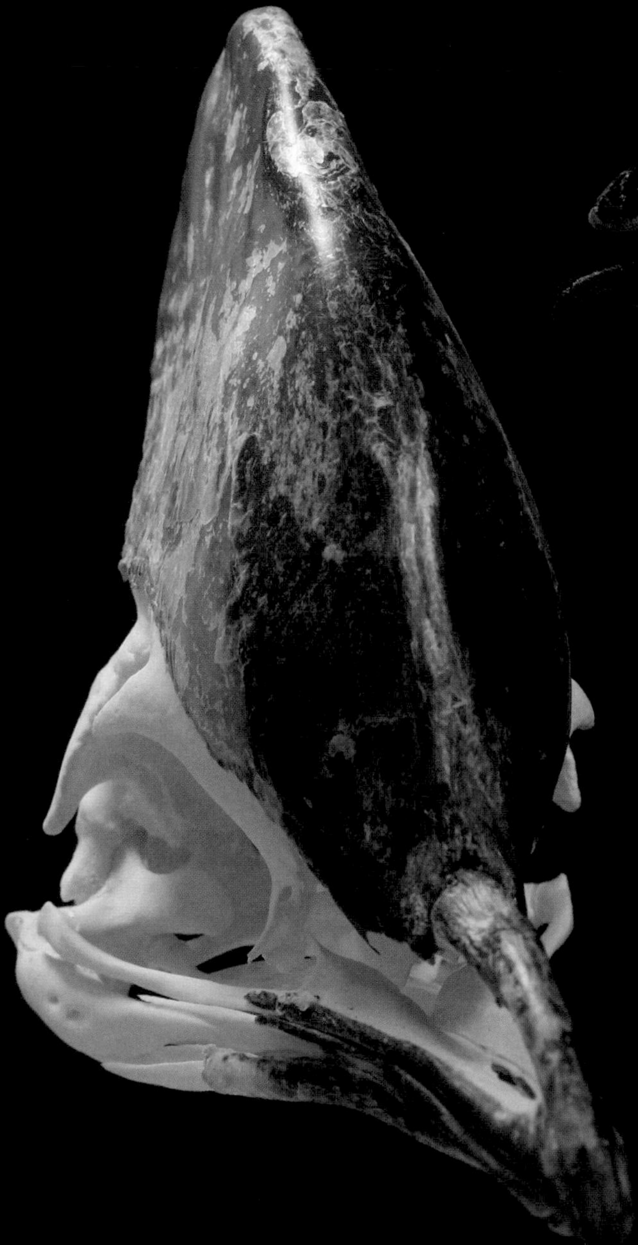

Southern Cassowary
Casuarius casuarius ◁

WHILE THE CASSOWARY'S beak is unremarkable, the skull does have a spectacular casque made of a firm keratinous shell that contains a soft interior. Cassowaries have elongated claws on their inner toes. They use these for defense as well as grubbing up insects.

Kingdom: Animalia
Phylum: Chordata
Class: Aves
Order: Struthioniformes
Family: Casuariidae
Genus: Casuarius
Behavior: Frugivore/
Diurnal

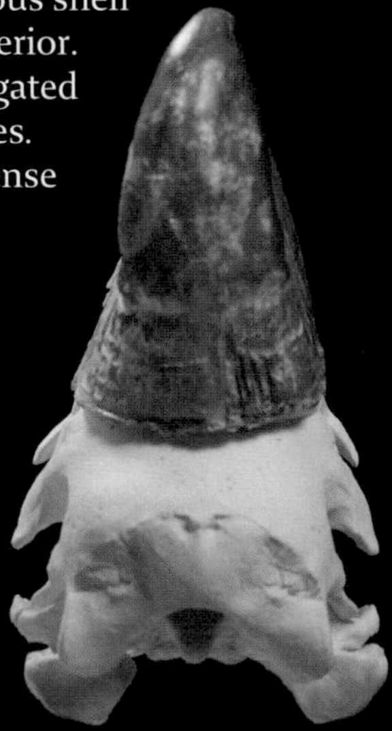

DUDLEY'S NOTES:
I have two cassowary skulls. These skulls are quite difficult to prepare because the bony, keratin cask on the top tends to come away from the skull. I do, however, like the honeycomb underneath the keratin cask.

Cassowaries, Ostriches, and Rheas

American Rhea
Rhea americana △

THE AMERICAN RHEA, the ostrich, and the cassowaries, among other birds, sit in a taxonomic clade named Palaeognathae. It was the features of the skull that taxonomists used to link these species together: they all share a common morphology of the palate with a broad vomer and a marked furrow on the bill just in front of the nostrils.

Kingdom: Animalia
Phylum: Chordata
Class: Aves
Order: Struthioniformes

Family: Rheidae
Genus: Rhea
Behavior: Omnivore/
Diurnal

Ostrich
Struthio camelus ◁

EYESIGHT IS EXTREMELY important to the ostriches, which inhabit semidesert areas of open land. Their large eyes combined with their great height allows them to spot potential predators from a considerable distance. Rather than digging their head in the sand, as myths suggest, ostriches flee at high speeds at the sight of a predator.

Kingdom: Animalia
Phylum: Chordata
Class: Aves
Order: Struthioniformes
Family: Struthionidae
Genus: Struthio
Behavior: Herbivore/Diurnal

Barn Owl
◁ *Tyto alba*

ALL OWLS ARE instantly recognizable, with their forward-pointing eyes, and characteristic flat facial disc of feathers. The large, round, circular bone that forms part of the eye socket (present only in the left eye of this skull) is known as the sclerotic ring. Owls have incredible vision, yet they can't move their eyes; instead, their necks can turn through almost 270 degrees. The eyes of owls typically occupy 30 to 50 percent of the skull volume. If humans were to have a similar relative eye size, we would have eyes as big as tennis balls.

Kingdom: Animalia
Phylum: Chordata
Class: Aves
Order: Strigiformes

Family: Tytonidae
Genus: Tyto
Behavior: Carnivore/
Crepuscular

European Eagle Owl
▷ *Bubo bubo*

INSTANTLY EVIDENT in this skull are the rings of bone that surround the owl's forward-facing eyes. These sclerotic rings compress the eyeball and prevent any movement so that the owl must turn its head to change its field of vision. The eagle owl sometimes hunts in daylight and is able to constrict its pupils to a pinprick to avoid damaging its light-sensitive retina. The wickedly hooked beak reflects the bird's carnivorous lifestyle: the eagle owl (the largest of the owls) is able to dispatch some formidable prey; for example, it can down small roe deer.

Kingdom: Animalia
Phylum: Chordata
Class: Aves
Order: Strigiformes
Family: Strigidae
Genus: Bubo
Behavior: Carnivore/
Nocturnal

Little Owl
▷ *Athene noctua*

The little owl (which really is little—measuring only twenty-one to twenty-three centimeters (9 inches) from top to toe) is predominantly a diurnal hunter, feeding on worms, amphibians, small mammals, and birds. Like its larger kin, it has wicked talons and a raptor's hooked beak. The skull is very delicate and dominated by the huge orbits and sclerotic rings.

Kingdom: Animalia
Phylum: Chordata
Class: Aves
Order: Strigiformes

Family: Strigidae
Genus: Athene
Behavior: Carnivore/
Crepuscular

DUDLEY'S NOTES:
I spotted this owl by the side of the road on my way home from work one day. I was amazed to find it was a little owl. It's quite common to find tawny owls, but much less so to find little owls.

Kea
Nestor notabilis △

THE KEA SKULL has a relatively large brain cavity, reflecting the bird's intelligence. A highly curious and sometimes cooperative bird, the kea uses its beak to gather its omnivorous diet. Some believe that one of the beak's more gruesome applications is to pierce through the wool and skin of sheep to feed on the subcutaneous fat of their backs; however, there is only anecdotal evidence for this.

Kingdom: Animalia
Phylum: Chordata
Class: Aves
Order: Psittaciformes
Family: Psittacidae
Genus: Nestor
Behavior: Omnivore/Diurnal

Budgerigar
Melopsittacus undulatus ▷

DESPITE ITS TINY SIZE, the budgerigar's beak is instantly recognizable as that of a parrot. Budgerigars are native to the drier parts of Australia but have found a home worldwide as popular pets. The longer upper mandible is used to de-husk seeds so that the nutritious kernel can be eaten.

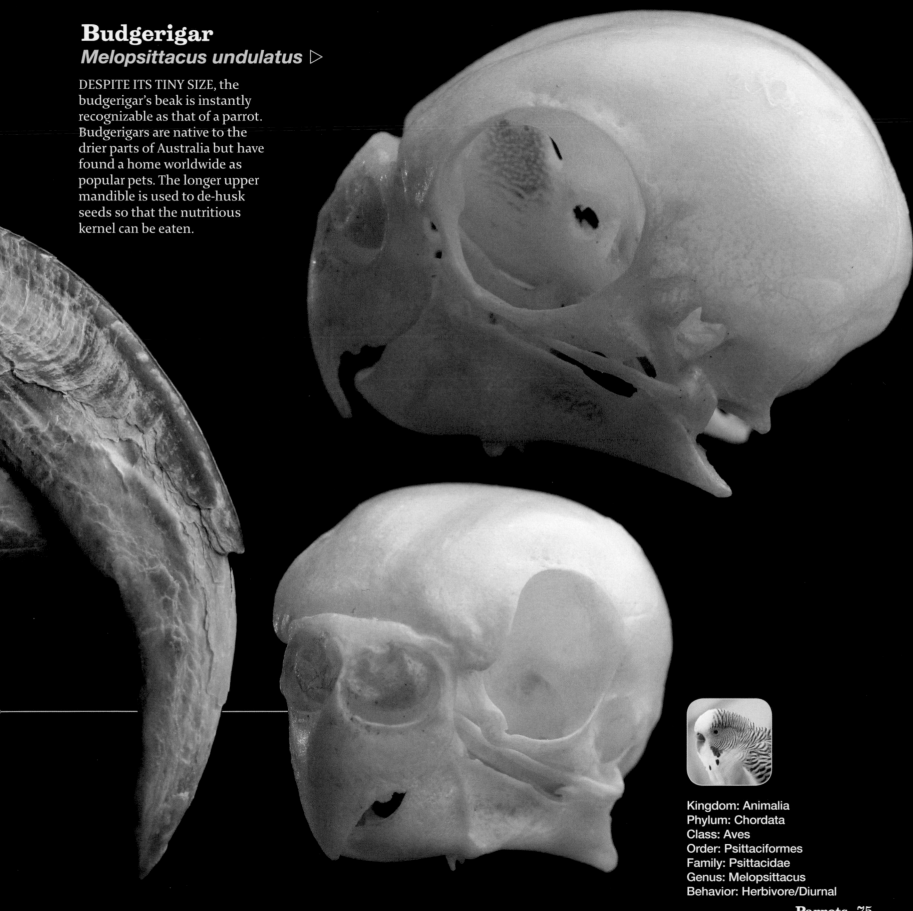

Kingdom: Animalia
Phylum: Chordata
Class: Aves
Order: Psittaciformes
Family: Psittacidae
Genus: Melopsittacus
Behavior: Herbivore/Diurnal

Hyacinth Macaw
▷ *Anodorhynchus hyacinthinus*

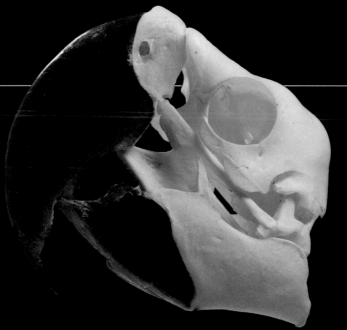

THE HYACINTH MACAW is the longest of all parrot species. Its large and powerful beak is used to crack open the shells of tough seeds such as macadamia and Brazil nuts. Macaws are observed first to score the shell with their upper mandible and then to use their strong jaw to cleave it neatly in two. Some nuts are too tough even for this bill, and macaws are known to sift through cattle dung to find partially digested acuri nuts to eat.

Kingdom: Animalia
Phylum: Chordata
Class: Aves
Order: Psittaciformes
Family: Psittacidae
Genus: Anodorhynchus
Behavior: Herbivore/ Diurnal

Red-and-Green Macaw
◁ *Ara chloropterus*

LIKE THE HYACINTH MACAW, this bird uses its powerful beak to break open seed cases. Several of the species of seed eaten by macaws contain toxic alkaloids that slowly build up in the birds. However, a cunning behavioral adaptation appears to have emerged to counter this: many species of macaw converge on banks in the Amazon in order to eat clay. It is thought that minerals in the clay have a detoxifying effect.

Kingdom: Animalia
Phylum: Chordata
Class: Aves
Order: Psittaciformes

Family: Psittacidae
Genus: Ara
Behavior: Herbivore/ Diurnal

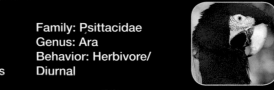

Pesquet's Parrot
▷ *Psittrichas fulgidus*

IN LIFE, THIS PARROT has a rather ugly, bald face reminiscent of that of a vulture. However, it is a frugivore, specializing in eating the sweet fruits of the fig family (and occasionally mangoes). Baldness is an adaptation to keep its feathers free of sticky residue.

Kingdom: Animalia
Phylum: Chordata
Class: Aves
Order: Psittaciformes

Family: Psittacidae
Genus: Psittrichas
Behavior: Frugivore/ Diurnal

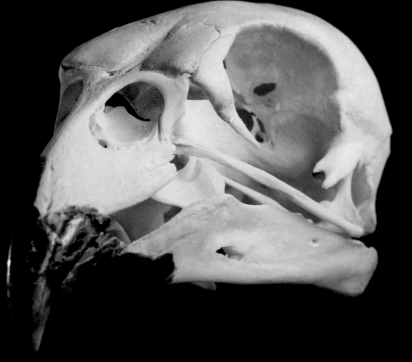

Lord Derby's Parakeet
Psittacula derbiana ▷

THE MALE LORD Derby's parakeet has a bright red beak with a yellow tip. This skull retains faint hints of its living coloration. These gregarious birds occupy upland coniferous forests in the eastern Himalayas and the Tibetan plateau.

Kingdom: Animalia
Phylum: Chordata
Class: Aves
Order: Psittaciformes
Family: Psittacidae
Genus: Psittacula
Behavior: Herbivore/Diurnal

Ring-necked Parakeet
Psittacula krameri

LIKE THOSE OF MANY other parrots in this collection, this species' enlarged upper mandible facilitates the dehusking, and cracking of hard cases on seeds; the serrated margin of the beak helps the bird manipulate fruit. Native to India and Africa, these noisy birds are now found in many parks around the world.

Kingdom: Animalia
Phylum: Chordata
Class: Aves
Order: Psittaciformes
Family: Psittacidae
Genus: Psittacula
Behavior: Herbivore/Diurnal

■ Passerines (Perching Birds)

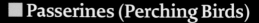

Carrion Crow
Corvus corone

CROWS ARE KNOWN for being remarkably intelligent. While the Caledonian crow takes the prize (competing with chimpanzees in their use, and construction of tools), the large brain-to-body size ratio of all corvids may explain the cosmopolitan success of the carrion crow. It is a general scavenger, eating carrion, worms, birds' eggs, and anything else it can get into its clutches; hence its rather generalized (but powerful) beak shape.

Kingdom: Animalia
Phylum: Chordata
Class: Aves
Order: Passeriformes
Family: Corvidae
Genus: Corvus
Behavior: Carnivore/
Diurnal

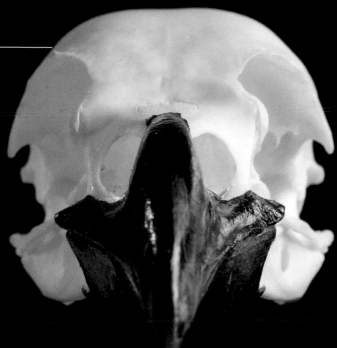

Common Raven
Corvus corax ◁

THE COMMON RAVEN is a scavenging bird (although ravens can also act as predators of small mammals in resource-poor regions) that inhabits most of the Northern Hemisphere. This species, in common with other members of the crow family, demonstrates exceptional intelligence and has one of the largest avian brains.

Kingdom: Animalia
Phylum: Chordata
Class: Aves
Order: Passeriformes
Family: Corvidae
Genus: Corvus
Behavior: Omnivore/
Diurnal

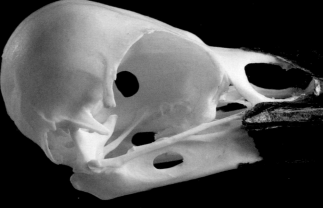

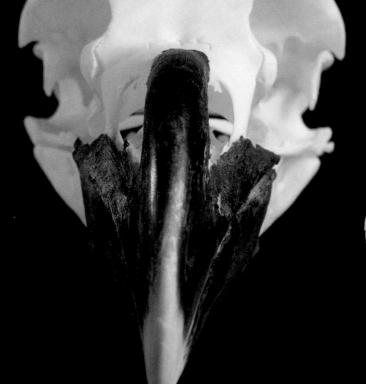

White-necked Raven
Corvus albicollis ◁

THIS AFRICAN RAVEN species is larger than the common raven (it has a one meter (3 foot) wingspan) but they have a similar diet: both are omnivorous and at times engage in opportunistic predation. The beak is suited to this lifestyle: powerful enough to crack seeds; hooked enough to take flesh from a carcass.

Kingdom: Animalia
Phylum: Chordata
Class: Aves
Order: Passeriformes
Family: Corvidae
Genus: Corvus
Behavior: Omnivore/
Diurnal

Common Crossbill
Loxia curvirostra

THE PECULIAR MANNER in which the upper and lower mandibles of this beak cross, instead of meeting, is an adaptation to the feeding behavior of these birds. They feed on seeds contained within pinecones and so ingeniously insert their beaks between the scales of their favorite variety to separate them and use their tongue to extract the seeds. Each crossbill species specializes on a small range of cone shapes.

Kingdom: Animalia
Phylum: Chordata
Class: Aves
Order: Passeriformes
Family: Fringillidae
Genus: Loxia
Behavior: Herbivore/Diurnal

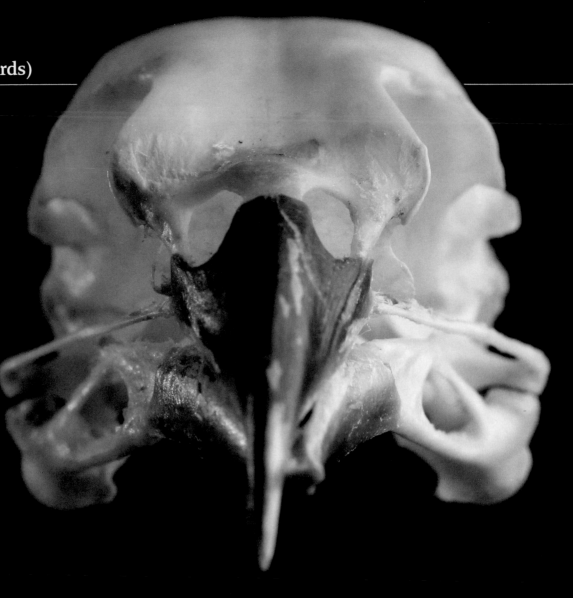

White-winged Grosbeak
Mycerobas carnipes

BELONGING TO THE true finches (*Fringillidae*), this bird has a typically short, thick beak and a robust skull that can withstand the large forces exerted by its strong muscles. These features allow the strong bite essential for dehusking tough seeds: the seed fits into a groove in the palate and is then crushed to release the nutritious kernel.

Kingdom: Animalia
Phylum: Chordata
Class: Aves
Order: Passeriformes
Family: Fringillidae
Genus: Mycerobas
Behavior: Herbivore/Diurnal

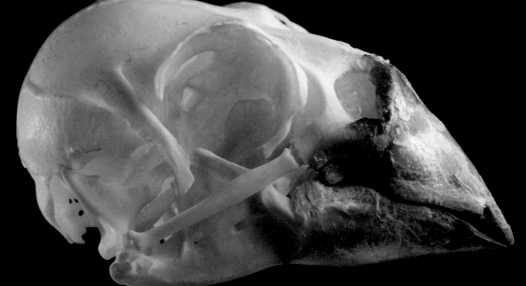

Spectacled Spiderhunter
Arachnothera flavigaster

THE LONG, AND SLENDER beak of this bird allows it to reach delicately and deeply into tubular flowers; its strawlike tongue then sucks up the nectar. Both parties benefit from this interaction; the spectacled spiderhunter's reward for carrying pollen to the next flower is a sip of sugar-rich solution. However, the species is named spiderhunter for a reason: members of the family are known to feast on arthropods as well.

Kingdom: Animalia
Phylum: Chordata
Class: Aves
Order: Passeriformes
Family: Nectariniida
Genus: Arachnothera
Behavior: Omnivore/Diurnal

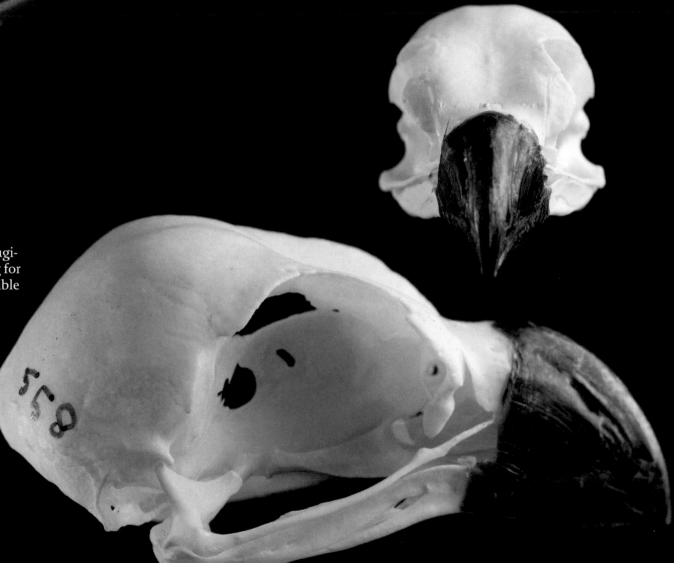

White-bellied Go-away-bird
Corythaixoides leucogaster

THIS AFRICAN BIRD is a frugivore with a particular liking for plantains. The upper mandible is serrated to facilitate the manipulation of fruits. They are named for their repeated *gwa* calls.

Kingdom: Animalia
Phylum: Chordata
Class: Aves
Order: Musophagiformes
Family: Musophagidae
Genus: Corythaixoides
Behavior: Frugivore/Diurnal

King Penguin
Aptenodytes patagonicus

THE KING PENGUIN has a long, slender bill suited to its lifestyle: this bird can regularly reach depths of more than 200 meters (650 feet) when on fishing voyages. The streamlined shape of its skull and bill mean more efficient and faster swimming, allowing these penguins to catch fish and squid.

Kingdom: Animalia
Phylum: Chordata
Class: Aves
Order: Sphenisciformes
Family: Spheniscidae
Genus: Aptenodytes
Behavior: Piscivore/
Diurnal

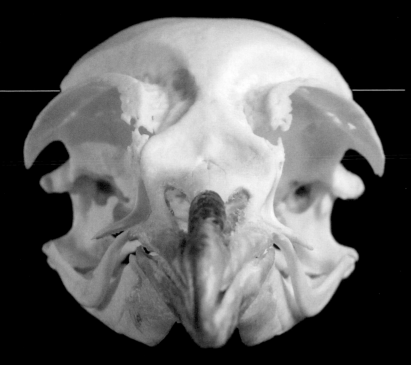

Gentoo Penguin
◁ *Pygoscelis papua*

THE GENTOO PENGUIN'S diet is made up predominantly of crustaceans such as krill. Their bills are heavier and less pointed than those of their fish-eating relatives.

Kingdom: Animalia
Phylum: Chordata
Class: Aves
Order: Sphenisciformes

Family: Spheniscidae
Genus: Pygoscelis
Behavior: Carnivore/Diurnal

Little Penguin
Eudyptula minor ▷

LITTLE PENGUINS are the smallest of all penguin species and sport a fetching slate blue coat of feathers. Living in burrows on the offshore islands of New Zealand and Australia, they hunt for fish and squid. As is typical of fishing penguins, they have a long, slender beak.

Kingdom: Animalia
Phylum: Chordata
Class: Aves
Order: Sphenisciformes

Family: Spheniscidae
Genus: Eudyptula
Behavior: Piscivore/
Diurnal

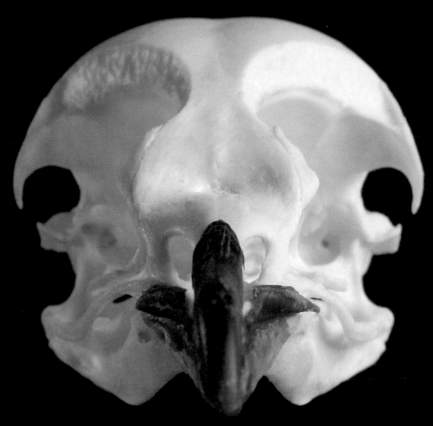

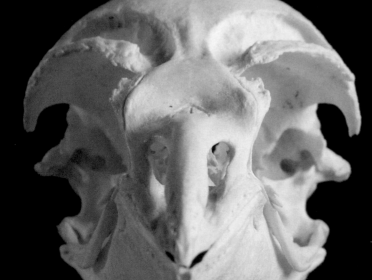

Southern Rockhopper Penguin
◁ *Eudyptes chrysocome*

LIKE THAT OF THE gentoo penguin, the southern rockhopper's beak is adapted for catching small, unmaneuverable crustaceans and is shorter and wider than the beaks of its faster fish-chasing cousins.

Kingdom: Animalia
Phylum: Chordata
Class: Aves
Order: Sphenisciformes

Family: Spheniscidae
Genus: Eudyptes
Behavior: Carnivore/
Diurnal

Blue Crowned Pigeon
Goura cristata

THIS IS ONE of the largest pigeon species, approaching the size of a turkey. It gets its name from the ornate crown of feathers that adorns the head of adult birds. The distinctly pigeonlike skull and beak morphology suit its frugivorous lifestyle. It is also the closest living relative to the dodo in the collection.

Kingdom: Animalia
Phylum: Chordata
Class: Aves
Order: Columbiformes
Family: Columbidae
Genus: Goura
Behavior: Herbivore/ Diurnal

Dodo
Raphus cucullatus

IT IS THOUGHT that the dodo used its heavy beak to feed mainly on fruits and seeds, fattening itself during the productive wet season to better survive the dry season on its native Mauritius. It is likely that this trait, combined with the overprovisioning of captive individuals, led to the caricature of the dodo as a fat, inelegant bird. However, much is speculation due to their untimely extinction, in the late seventeenth century, at the hands of invasive predators (dogs and pigs, we now think, rather than hungry sailors).

Kingdom: Animalia
Phylum: Chordata
Class: Aves
Order: Columbiformes
Family: Raphidae
Genus: Raphus
Behavior: Herbivore/Diurnal

The Skull of the Dodo

EUROPEANS KNEW THIS remarkable bird for little more than eighty years, a span so brief as to offer an especially melancholy commentary on humankind's relationship with the natural world—for we were the ones responsible for this creature's extinction.

Dutch sailors blown off course en route to the East Indies in 1598 were the first to settle on the island that they named Mauritius in honor of Prince Maurice of Orange. There they saw and wrote about the unusual-looking but locally quite common bird, of a type unknown elsewhere, nesting among the trees and the tussock grass. Large-headed, large-billed, and ungainly, it was a forty-pound creature with wings so tiny as to render it quite unable to fly. The Portuguese, who landed briefly on the island some while before, had called it a doudo, their vernacular for a simpleton, a fool. The Dutch saw no reason to give the creature any other name: it appeared as stupid to them as it had to their southern European cousins. They called it the dodo .

This giant bird, which when startled could flap no more than a few inches into

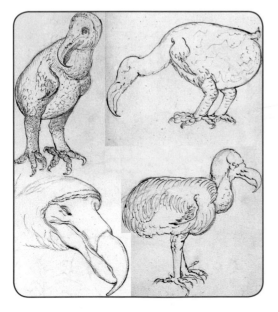

Some of the first depictions of dodos on the island of Mauritius, made during the voyage of the VOC Gelderland in 1602.

the air as it ran away, turned out to be ludicrously easy to hunt, like a turkey—but unlike a turkey its meat was tough and tasted awful, and so the Dutch left it alone except for occasional sport. But with the Europeans came also the inevitable ship-borne menagerie that has wrought so much environmental damage elsewhere, the dogs, cats, pigs—and rats. As these beasts expanded into the meadows of Mauritius they discovered the nests of the amiable dodos and attacked and killed them with pathetic ease and impunity.

Halfway through the seventeenth century some mildly alarmed Dutchmen, who were by then starting to show more than typical colonial solicitude for the local flora and fauna, started to protest the very evidently declining fortunes of this sad old bird. But to little avail. By 1675 the population of dodos on Mauritius was reduced to a couple of breeding pairs; by 1680 the rats had claimed the last living specimen. *Didus ineptus*, initially and oh-so-appropriately assigned to the family Colombidae, was no more.

In not much more than eight decades after European colonialists first set eyes on the bird, their carelessness—or really, our carelessness, since it's now acknowledged that few of us can claim ancestors who did much better—had rendered this entire species of ill-favored and flightless creatures extinct. It would not be the last extinction witnessed by modern man (think some tigers, great auks) and was probably not the first; but the dodo died before our very eyes, and its passing shames us still, right down to the present day.

We now have a better idea, incidentally, of what exactly the dodo was. Eighteenth-century biologists were right to classify it among the Colombidae, since it clearly is a kind of pigeon or dove. Close study of

its bones and skull, even back in the 1700s, showed that. But pigeons and doves can fly. Recently, to accommodate those that can not, avian biologists created a new subfamily for flightless pigeons, naming them the Raphinae. Within this group so far are two bird species—a tall ibis-like creature known as the *Rodrigues solitaire*, and our dodo, now named *raphus cucullatus*. Neither could fly; both are now extinct. Gene sequencing has shown that another Indian Ocean bird, the Nicobar pigeon, which is also is large but can fly, is the Raphinae's closest living relative.

As to why the dodo was so large: A theory advanced in the 1960s by the evolutionary biologist J. Bristol Foster holds that, on small islands particularly, certain species become larger or smaller depending on the resources available locally and on the absence or presence of predators. The dodo had an abundant supply of fruit and few worries from competing creatures—and so it became ever fatter and fatter. It needs to be stressed that it did not become extinct as a direct result of its obesity, but from the importation of predators.

The odd appearance of the dodo made it something of a celebrity bird among seventeenth-century swells, causing a number to be captured and brought back to Europe. Two Dutch painters, Roleant Savery and his nephew Jan, became well known for painting these captive specimens. The fatness of the dodos depicted by the pair is now thought to be something of a distortion; their live models had gorged themselves on fruit given to them by their captors. "Slimline" images of dodos, presumably more faithful to their appearance in the wild, are now becoming rather more common.

Once the living specimens dwindled and vanished, skeletons of dodos, or the

The dodo died before our very eyes, and its passing shames us still, right down to the present day.

stuffed versions fashioned by taxidermists in London and Amsterdam, became popular and much favored by collectors. One such collector was John Tradescant, a prominent gardener and silkworm expert who in the early seventeenth century maintained, in Lambeth, a popular "cabinet of curiosities"—the first proper museum in Britain. On his death his collection passed to Elias Ashmole, who was himself creating a museum in Oxford. Among the items in the twelve wagon loads of exotic materials handed over to the safe-keeping of Oxford University was a stuffed dodo.

It seems to have been stuffed inexpertly, however, and soon began to rot away most disagreeably. By 1755 it was in so lamentable a condition that the Ashmolean Museum's then director ordered it to be removed from display and dispatched into the attic. By the time the Oxford Museum of Natural History was built a century later, and expressed an interest in having what remained, only the mummified skull and a skeletal foot were left—comprising not just the skull and bones, but also, it turned out, the only soft tissue of a dodo remaining anywhere in the world.

These rather pathetic remains, together with paintings by the Saverys, form the centerpiece of what is now the most valued dodo collection in the world—a collection that, since it includes that soft tissue, is a source of DNA for all manner of future research into this curious, romantic, and now long-lost bird.

The romance, such as it is, derives largely from interest shown in the dodo's early appearance in the Oxford museum by a scholar-neighbor, Charles Dodgson, who had been taking calotype photographs of its collections in the 1860s. Dodgson was of course the creator of *Alice in Wonderland*, Lewis Carroll. One of Dodgson's many peculiarities was his stutter—an affliction that had long made him incapable of properly saying his own given name—he would introduce himself as Charles Do-Do-

Dodgson. Because of this, he explained, he felt a special affinity with this strange bird whose name, the dodo, was so similar to his distorted own.

And thus, it is now thought, Lewis Carroll included the dodo as a character in Alice—essentially as a self-portrait. The book's first and arguably finest illustrator, John Tenniel, then drew pictures of the bird, based on the amusingly chubby versions that had been created by the Saverys, uncle and nephew, two centuries before.

The dodo thus acquired its popular fame—a fame that has clung ever since to this tragically departed creature, an emblem of the clumsy carelessness of humankind, and of the fragility of nature, into which man has been wont so casually to intrude.

And the Oxford dodo in particular, though now merely a skull with just a tiny amount of adhering flesh, kept secure beside its remaining foot in a glass case, stands as a poignant symbol of a loss of human innocence, but also of our wonder at a fellow creature now eternally gone from the natural world; gone, but thanks to an enduring classic of children's literature, never likely to be forgotten.

This painting, one of the most famous and often copied representations of a dodo, was painted by Roleant Savery in 1626 and is said to have inspired John Tenniel.

A previously unpublished illustration of a dodo specimen from the seventeenth century. The inscription, which reads "Dronte," was the seventeenth-century Dutch name for the dodo.

Black Crowned Crane
Balearica pavonina

THE BLACK CROWNED CRANE is one of the most primitive living cranes. In many other species of crane, the trachea is long and coils into the sternum; this extension to the trachea give these birds the ability to call with a loud, far-carrying bugle. By contrast, the black crowned crane lacks coils in the trachea and, consequently, produces considerably less regal honking sounds. It can also produce a booming sound by inflating and then pushing air out of the small red gular pouch below its bill.

AKA: Kaffir Crane
Kingdom: Animalia
Phylum: Chordata
Class: Aves
Order: Gruiformes
Family: Gruidae
Genus: Balearica
Behavior: Omnivore/Diurnal

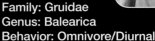

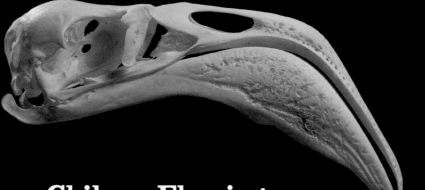

Chilean Flamingo
Phoenicopterus chilensis △

FLAMINGOS HAVE CURIOUS beaks that are used upside down when the bird dips its head into water to sift through silt for blue-green algae and shrimp. It is this diet that gives them their famous pink color. A row of hairlike structures helps trap these small food items as the beak passes through the water.

Kingdom: Animalia
Phylum: Chordata
Class: Aves
Order: Ciconiiformes
Family: Phoenicopteridae
Genus: Phoenicopterus
Behavior: Omnivore/Diurnal

Lesser Flamingo
Phoenicopterus minor △

THE CHARACTERISTIC PINKISH color of the lesser flamingo is due to its diet. The filtering system of hair-like structures at the margins of the mandibles has been wonderfully preserved in this specimen. When feeding, the head is deployed upside down and is swept through the water in side-to-side movements. Lesser flamingos gather in huge numbers on the lakes of Africa's Rift Valley.

Kingdom: Animalia
Phylum: Chordata
Class: Aves
Order: Ciconiiformes
Family: Phoenicopteridae
Genus: Phoenicopterus
Behavior: Omnivore/ Diurnal

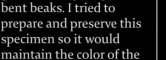

> **DUDLEY'S NOTES:**
> What I like about flamingo skulls is that they've got unusually shaped, bent beaks. I tried to prepare and preserve this specimen so it would maintain the color of the beak.

Eurasian Spoonbill
Platalea leucorodia ▽

THE APTLY NAMED spoonbill wades through shallow water and moves the flattened end of its bill from side to side through the water and mud to disturb prey (such as invertebrates, tadpoles, and fish) from their hiding places and quickly snatch them.

AKA: Spoonbill
Kingdom: Animalia
Phylum: Chordata

Class: Aves
Order: Ciconiiformes
Family: Threskiornithidae
Genus: Platalea
Behavior: Carnivore/ Diurnal

Shoebill
Balaeniceps rex ▷

THIS IS PERHAPS the most beautiful bird skull in the collection. The shoebill's extraordinary bill retains some of its natural coloring, and the sharp hook at the end is obvious. These east African storks prey on fish and amphibia in muddy water. It is worth comparing this skull with that of the boat-billed heron.

AKA: Whalehead
Kingdom: Animalia
Phylum: Chordata
Class: Aves
Order: Ciconiiformes
Family: Pelecanidae
Genus: Balaeniceps
Behavior: Carnivore/Diurnal

DUDLEY'S NOTES:
The shoebill is the most sought-after bird skull in the world and I don't know another collector who has one other than myself. It looks like a clog with a big hook on the end. I was lucky to get this one and even luckier that it has retained so much of its coloring.

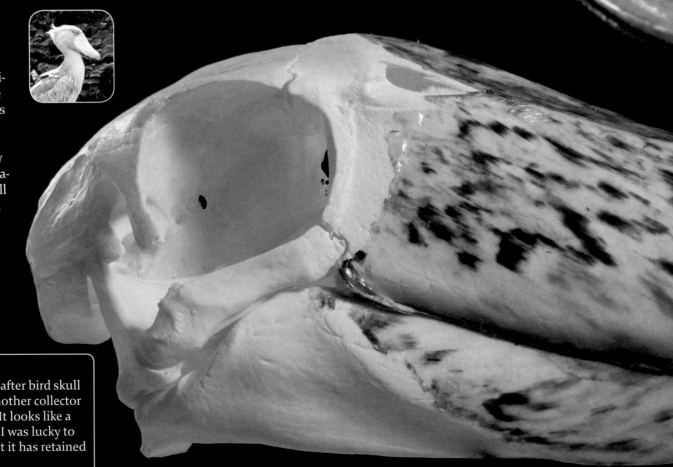

Boat-billed Heron
Cochlearius cochlearius ▷

THE BILL OF THE boat-billed heron is remarkably heavy and broad. Reference to the bill is found in its scientific name, which is derived from the Latin *cochlearium* or "spoon." These herons are found in the mangrove swamps of South America.

AKA: Boatbill
Kingdom: Animalia
Phylum: Chordata
Class: Aves
Order: Ciconiiformes
Family: Ardeidae
Genus: Cochlearius
Behavior: Carnivore/Diurnal

DUDLEY'S NOTES:
The boat-billed heron is one of my favorite skulls. The skull itself looks almost as if it's a mistake of nature or as if somebody has stepped on the beak and flattened it.

Hadada Ibis
Bostrychia hagedash ◁

The hadada ibis uses its scimitar-shaped bill to probe lawns and grassy areas for invertebrates. These birds are named for their loud *haa-haa-de-daa* calls.

AKA: Hadeda Ibis
Kingdom: Animalia
Phylum: Chordata
Class: Aves
Order: Ciconiiformes
Family: Threskiornithidae
Genus: Bostrychia
Behavior: Carnivore/Diurnal

Asian Openbill
Anastomus oscitans ▷

AS YOU CAN SEE from the skull, these storks get their name from the fact that their upper and lower mandibles do not close completely except at the very tip of the beak, due to a depression in the lower mandible. The gap serves to help the birds grip the freshwater snails that make up most of their diet.

AKA: Open-billed Stork
Kingdom: Animalia
Phylum: Chordata
Class: Aves
Order: Ciconiiformes

Family: Ciconiidae
Genus: Anastomus
Behavior: Carnivore/
Diurnal

Jabiru
Ephippiorhynchus mycteria ▷

THE JABIRU STABS its long, hefty bill quickly and repeatedly into relatively deep water to flush its prey from hiding. These birds are enormous, reaching up to 140 centimeters (55 inches) in height; they have been observed tackling prey as large as young caiman, which they drag back to shore to dismember before eating.

Kingdom: Animalia
Phylum: Chordata
Class: Aves
Order: Ciconiiformes

Family: Ciconiidae
Genus: Ephippiorhynchus
Behavior: Carnivore/
Diurnal

Maguari Stork
Ciconia maguari ▷

THE MAGUARI STORK is widespread throughout most of South America east of the Andes. Typical of its genus, it has a long, dagger-shaped bill over twice the length of its head. This is well suited to its favored feeding method, which involves picking among dense aquatic vegetation in shallow waters (as well as agricultural fields) in search of amphibians, fish, crustaceans, insects, and some rodents.

Kingdom: Animalia
Phylum: Chordata
Class: Aves
Order: Ciconiiformes

Family: Ciconiidae
Genus: Ciconia
Behavior: Carnivore/
Diurnal

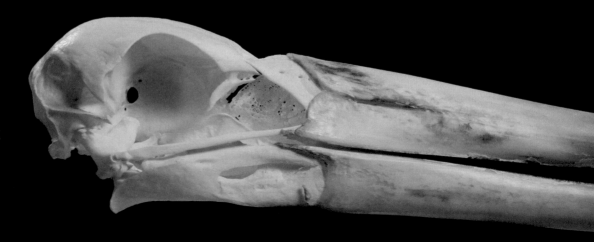

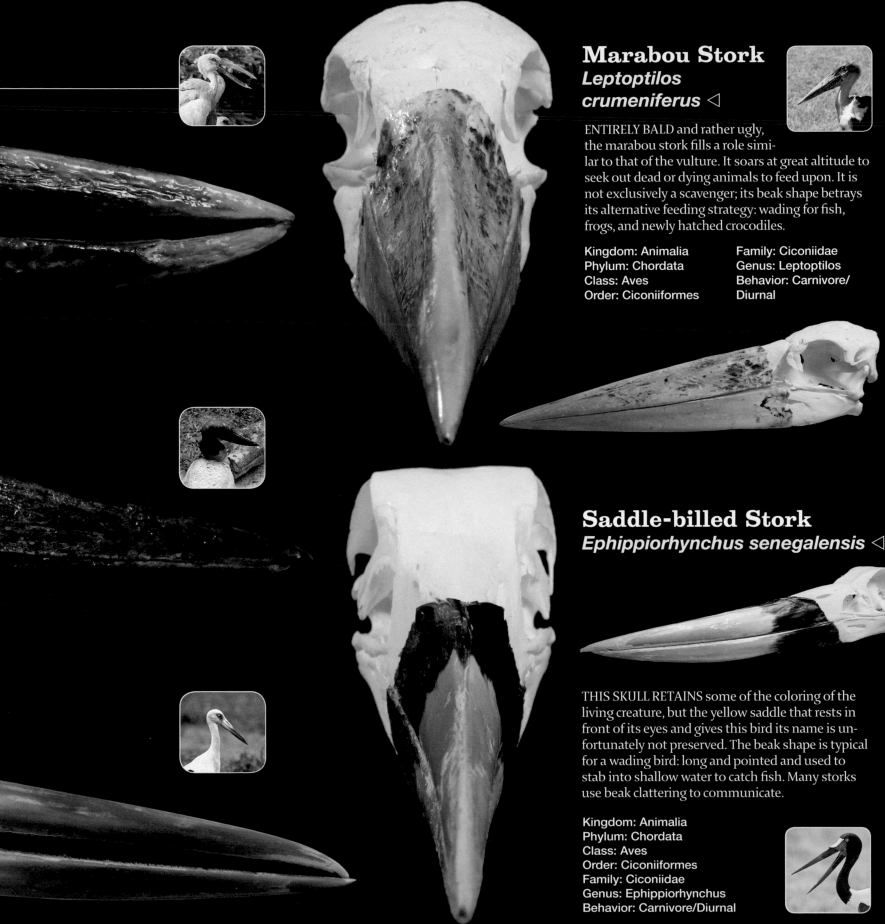

Marabou Stork
Leptoptilos crumeniferus ◁

ENTIRELY BALD and rather ugly, the marabou stork fills a role similar to that of the vulture. It soars at great altitude to seek out dead or dying animals to feed upon. It is not exclusively a scavenger; its beak shape betrays its alternative feeding strategy: wading for fish, frogs, and newly hatched crocodiles.

Kingdom: Animalia
Phylum: Chordata
Class: Aves
Order: Ciconiiformes
Family: Ciconiidae
Genus: Leptoptilos
Behavior: Carnivore/ Diurnal

Saddle-billed Stork
Ephippiorhynchus senegalensis ◁

THIS SKULL RETAINS some of the coloring of the living creature, but the yellow saddle that rests in front of its eyes and gives this bird its name is unfortunately not preserved. The beak shape is typical for a wading bird: long and pointed and used to stab into shallow water to catch fish. Many storks use beak clattering to communicate.

Kingdom: Animalia
Phylum: Chordata
Class: Aves
Order: Ciconiiformes
Family: Ciconiidae
Genus: Ephippiorhynchus
Behavior: Carnivore/Diurnal

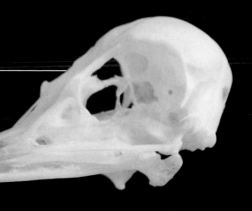

Pied Avocet
Recurvirostra avosetta

THE PIED AVOCET is a striking bird familiar to British people as the logo of the Royal Society for the Protection of Birds (RSPB). The upturned bill (the recurvirostra of its scientific name) is swiped from side to side on the surface of mudflats, searching for worms, crustaceans, and small fish.

AKA: Black-capped Avocet
Kingdom: Animalia
Phylum: Chordata
Class: Aves
Order: Ciconiiformes
Family: Charadriidae
Genus: Recurvirostra
Behavior: Insectivore/Diurnal

DUDLEY'S NOTES:
I love these birds because of the upturned bill. Very few birds have this. The avocet's skull is also very delicate. This one was killed at the zoo by a stoat. It was quite tragic.

White Stork
Ciconia ciconia

AS MANY STORKS do, this one communicates primarily by clattering its beak. The long red beak is also used to capture a variety of prey species (including large insects, frogs, and reptiles); it walks through low vegetation and shallow water with its beak poised ready to strike. In Europe it is considered auspicious if a white stork nests on your roof.

Kingdom: Animalia
Phylum: Chordata
Class: Aves
Order: Ciconiiformes
Family: Ciconiidae
Genus: Ciconia
Behavior: Carnivore/Diurnal

Great Northern Diver
Gavia immer ▷

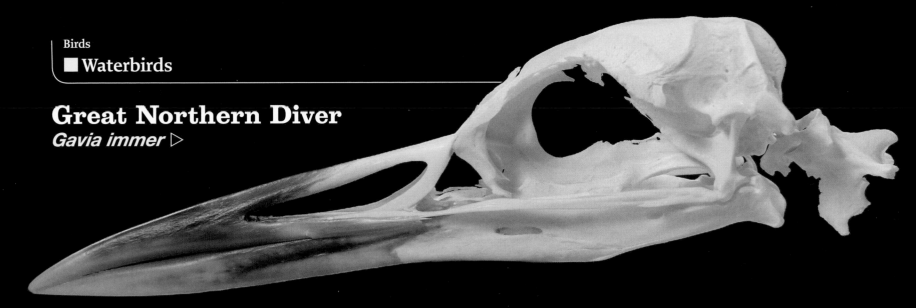

THIS FISHING SPECIALIST can dive as deep as an incredible sixty meters (200 feet) in search of prey with its long, sharp beak. The beak is also employed in chasing away nest predators or larger birds seeking to steal the diver's catches. The haunting mating calls of the great northern diver give rise to its American name of loon.

AKA: Common Loon
Kingdom: Animalia
Phylum: Chordata
Class: Aves
Order: Ciconiiformes

Family: Gaviidae
Genus: Gavia
Behavior: Piscivore/ Diurnal

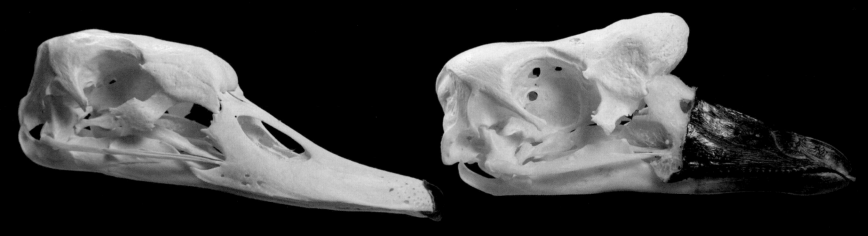

Mute Swan
Cygnus olor △

THE MUTE SWAN is familiar to many of us as an ornamental bird of parks and gardens. The black area at the tip of the bill is known as the nail and is a common feature of many ducks, geese, and swans.

Kingdom: Animalia
Phylum: Chordata
Class: Aves
Order: Anseriformes
Family: Anatidae

Genus: Cygnus
Behavior: Herbivore/ Diurnal

Chinese Goose
Anser cygnoides △

THIS DOMESTICATED GOOSE species (closely related to the swan goose) has a peculiar and obvious knob on its beak, more pronounced in males than females. The Chinese goose is a particularly large and noisy species, so it is often used as a "guard dog" by free-range poultry farmers.

Kingdom: Animalia
Phylum: Chordata
Class: Aves
Order: Anseriformes

Family: Anatidae
Genus: Anser
Behavior: Herbivore/ Diurnal

Surf Scoter
▷ *Melanitta perspicillata*

RATHER THAN SIEVING the surface, as the northern shoveler does, this duck dives for crustaceans and has a more bulbous bill than its dabbling relatives. The male's bill is brightly colored in the breeding season.

Kingdom: Animalia
Phylum: Chordata
Class: Aves
Order: Anseriformes

Family: Anatidae
Genus: Melanitta
Behavior: Diurnal

Northern Shoveler
▷ *Anas clypeata*

THE NORTHERN SHOVELER sweeps its wide bill across the water's surface, the hairlike structures at the edge of the bill sieving out small invertebrates. This filtering process is an example of morphological convergence across a great taxonomic divide, ranging from these tiny hairs to the giant and elaborate sieves of the baleen whales.

AKA: Shoveler
Kingdom: Animalia
Phylum: Chordata
Class: Aves
Order: Anseriformes
Family: Anatidae
Genus: Anas
Behavior: Omnivore/
Diurnal

Great Crested Grebe
▷ *Podiceps cristatus*

THE SLENDER, pointed beak of this species is an adaptation to its diet of fish and crustaceans. It was hunted to near-extinction in Britain as its head plumes (or aigrettes) were much sought after to adorn ladies' hats. These grebes have staged a remarkable recovery since a group of women got together to make a stand against the wearing of feathers.

Kingdom: Animalia
Phylum: Chordata
Class: Aves
Order: Ciconiiformes
Family: Podicipedidae
Genus: Podiceps
Behavior: Piscivore/Diurnal

Great White Pelican
◁ *Pelecanus onocrotalus*

THIS EXTRAORDINARY SKULL has a bill more than thirty centimeters (1 foot) long. Great white pelicans catch fish with a huge gular sac held open in the form of a scoop. They are gregarious birds and sometimes fish cooperatively in groups, by forming a semicircle with the open side forward and scooping up fish with rapid dips of the bill into the water.

AKA: Eastern White Pelican
Kingdom: Animalia
Phylum: Chordata
Class: Aves
Order: Ciconiiformes
Family: Pelecanidae
Genus: Pelecanus
Behavior: Piscivore/Diurnal

Science and Pseudoscience

I T GOES ALMOST without saying that the skull, as an entity, has been entirely fascinating to the scientist—or at least it has since that word, ending in the same-*ist* as in *artist*, *atheist*, and *economist*, was first used in the mid-1830s, at the beginning of Queen Victoria's reign. And indeed, why should the hard part of the head not attract the attentions of those particular men and women who are gripped with a hunger for knowledge?

For the skull contains the centerpiece of an animal's consciousness. Within its bony protection is the creature's command center, its mission control, its operations room, its executive office. It is where the major sense organs are located, the place where all the signals from those organs—be they ears or noses, eyes or tongues—are received, processed, and acted upon. So why would not this fantastic lode of life's most valued riches—which include not just the ability to move, or hear, or see, but also the unquantifiable mysteries of the emotional world, of love, of intellect, of memory, of fear—why would this not have become a mine of limitless extent for those whose business is to seek out the truths of the natural world? The skull is just so obvious a center for study that its fascination goes almost without saying.

As, however, does something else. Precisely because the skull houses so much, because it is so recognizably, and literally, the body-less embodiment of all advanced life, it attracts not just scientists but also the polar opposite of scientists, and in immense numbers. It attracts, in other words, pseudoscientists.

Whether it is of a human, a humanoid, a fish, a dinosaur, or a goat, the skull has become the locus of a great number of concerns that have little or nothing to do with science, in the strict sense: skulls have become central to ritual and myth, to legend and rite, to worship and sacred imagery, to irrational fear, to cults and the occult. It is in fact a curious irony: the part of the body that finds itself subjected to so much rigorously dispassionate inquiry is also the part that has lent itself to centuries of utterly irrational belief.

Scientific method is based on firm principles, established over the years and generally observed with great zeal. The process starts with a predicament, a question. Why, to take a very simple example, does a walrus skull have long downward-pointing tusks, when most other animals with tusks—elephants, boars, peccaries—sport them pointing upward?

To answer this, the scientific method demands first an hypothesis: the construction of a reasonable idea, or a series of reasonable ideas, that might explain why the walrus is so equipped. You might hazard, as one idea, the notion that the walrus uses its tusks to forage on the sea-floor; or then, as another, that it employs them in place of fins or arms to haul itself out of the water and back onto the ice; or thirdly, that its tusks are forms of sexual display, deployed with the specific purpose of finding a mate.

To test these hypotheses, you then observe. You watch walruses by the hundreds, in various places and at various seasons, and you take notes. Each time you suspect that a male walrus is flaunting his tusks to attract a female, you note it; each time he hooks his tusks onto the edge of an ice floe and hauls himself up and out of the water, you note it; and each time you see him diving and returning to the surface with his tusks covered with mud and weed, and the walrus himself with a mouth stuffed with food, you note it. And then at the end of the observing period the collected figures present a tempting confirmation of one or more of the three hypotheses—in this case, with the greatest number of notes relating to the times that the tusks enabled the animal, weary from swimming, to lever itself out onto an ice floe for a period of rest. This most popular notion—that a principal use of walrus tusks is for hauling itself out of the water—then steps up a notch, from being an hypothesis to becoming a theory.

A theory is an idea that can then be tested. Experiments can be conducted to try out the theory's value, its strength—let us say, cruelly, by removing a walrus's tusks and seeing just how it fares when trying to leave the water. If the tuskless walrus finds it impossible to get itself out of the water, and if, moreover, this same experiment repeated time and again shows exactly the same result—no tusks, no escape—then the theory is transformed into almost a fact, almost an item of knowledge. All that then remains is to disseminate the idea among all those other scientists working in the same field of study, to see whether their own research and observation confirm or refute what you claim to have discovered.

Let us say all this is duly confirmed. You are now able to say, definitively, just why a walrus sports its downward-pointing tusks—and at the same time you can now rule out one of the many possible reasons that an elephant's upward-pointing tusks have such a very different configuration. One can then start to construct a family of elephant tusk hypotheses, empirical data can be collected from the systematic observation of elephant behavior, an elephant tusk theory can be adduced, it can be tested by experiment, and a conclusion reached here also, and so on and so on.

And thus, by the often tedious eight-stage rigors of the scientific method—(1) the posing of a question, (2) the creation of hypotheses, (3) the collection of empirical data from observation that allows a ranking of these various competing hypotheses, (4) the crucial adducing of a theory, (5) the collection of empirical data from experiment to confirm or refute the theory, (6) the discovery of the theory's near-proof, (7) the final polish given to this near-proof by its review and evaluation by peers, and (8) the final publication of the proof—by this method does scientific knowledge advance and human ignorance and prejudice slowly recede.

In the case of the skull, so fascinating an object, this is a process that is, and has been, repeated countless times. Why does the toucan have such a colorful beak? How could a saber-tooth cat possibly wield such massive canine teeth? Why does a rabbit skull have such immense auditory bullae? Is there a connection between the size of sagittal ridges and the eating habits of animals that possess them? In each case, question-hypothesis-observation-theory-experiment-proof, question-hypothesis-observation-theory-experiment-proof, and through this relentless process we have learned things about skulls and have come to know exactly why one looks a certain way or was built in a particular fashion. Or else our method hasn't offered a ready answer at all, and we puzzle still. Why does a narwhal's tusk never grow out of the right side of its jaw? We've wondered, suggested, asked, experimented—but we still have no idea.

Just the opposite of this process goes on with pseudoscience—with the result that its practitioners seldom utter, "We have no idea." For it's axiomatic that pseudoscientists, especially those whose business is the skull, always have ideas, and ideas that, at least for a while (for much in pseudoscience is merely of the moment), are held very firmly indeed.

A classic example, and one especially relevant here, is that of craniometry, the measurement of the skull, and one particular pseudoscientific idea that such measurements once generated: the idea that humankind could be classed into a number of basic races, and that these races could be placed in a definite hierarchy according to what was called the cephalic index, which measured and rated skull shapes.

A happily now discredited group of eugenicists proposed that, based on this index, humankind could be shown to range from the superior, dolichocephalic races (the Aryans) with their relatively long and narrow skulls, to the brachycephalic peoples, whose broad and squat skulls indicated, according to the contemporary literature, that they were "mediocre and inert peoples...best represented by the Jew." Not surprisingly,

The adaptive significance of the walrus tusk, unique among marine mammals, has been attributed to many different factors. Similar to elephant tusks, they appear to be multifunctional.

these "results", which were first bruited at the turn of the twentieth century, were seized upon by a variety of authoritarian governments, the Nazis in Germany first, the white South Africans later, to justify race-based governance as well as, notoriously, pogroms and exterminations.

There is a basis of science here, of course. No one disputes the validity or the value of the systematic measurement of skulls (we gaily classify dogs and cats on this basis: a Boston terrier is brachycephalic, a Greyhound dolichocephalic); nor is there any real argument that different racial types have generally different skull types. But where the science becomes perverted by the pseudoscientific method is at the point of interpretation: the sudden assertion—based on no evidence, no proven hypothesis, and no tested theory—that these different racial types are to be ranged into some kind of hierarchy, with one type superior to another in intelligence, ability, or moral tenor.

Invariably trouble follows from such a monstrous claim: for those who reach these conclusions generally like, perversely, to publish and disseminate their "findings" widely—despite being well aware of the

social implications that such wildly distorted information inevitably invites. Aryans are "superior," for instance, is an assertion, wholly untrue, that has in recent memory led to the deaths of millions, as well as to the worldwide spread of abject misery. And all of it stemming in this case from the measurement of skulls, and the misdirection of the scientific mind by perverse men of ill will.

Much the same kind of departure from scientific rigor—though it is not usually hijacked for political motives, as Hitler's henchmen so notoriously did with craniometric data in the 1920s—is what characterizes almost all pseudoscience. Whether this "science" is homeopathy or reiki, astrology or alchemy, telepathy or channeling, there's invariably just enough science to be found to convince the gullible, yet never enough to offer a truly convincing proof. All scientific endeavor is tainted by such dishonest inference, such falsity through partial truth—whether in the study of humankind's origins, or of individuals' ailments, or of their character and strength of moral fiber. In all of these—all here related, one way or another, to the skull—there is the science; and yet more ominously, there's also the pseudoscience, quietly mocking from the wings, yet steadily gaining converts all the while.

Piltdown Man

CHARLES DAWSON, a Victorian solicitor with a practice in Sussex, liked to be thought of as very much the gentleman—his official portrait shows him standing in a drawing room with a Doric column, wearing velvet breeches and a frock coat and sporting a tricorn hat and a sword. And he was not to be thought of as simply a gentleman, but as a gentleman-scholar: he was an amateur student of archaeology and an authority on geology, and had a jumble of letters next to his name to prove it.

He was, moreover, insatiably ambitious, determined to win a worldwide reputation—and it was in this endeavor he seems to have overreached himself, because all evidence now suggests that it was this Mr. Dawson who perpetrated on the world of science one of the longest-lasting and most infamous frauds of all time. Charles Dawson, it is now firmly believed, was the creator of Piltdown Man.

It was in 1908 that the then forty-four-year-old Dawson found, by his own account, the first of a number of fragments of an apparently very ancient and very unusual human skull. Over the next four years, as he searched the Pleistocene gravels in a quarry pit close to Barkham Manor, a grand country house in the Sussex village of Piltdown, he came across still more pieces—the quarrymen seemingly having found a complete skull but having smashed it to pieces, for reasons best known to themselves. The skull looked, to Dawson's half-expert eye, to be most odd—its features suggesting it might be half man, half monkey.

Dawson, his excitement mounting as his finds began to confirm these suspicions, alerted Arthur Smith Woodward, the head of the geological department of the British Museum and one of the greatest geological minds in Britain. Woodward came down to Sussex, the two searched the site together,

and they came across yet more pieces—the recognizable smooth parts of a cranium, half of a mandible (the lower jaw bone), and two molar teeth. Woodward took them all back to London with him, and began a scrupulous examination.

At the time, the British scientific community was in the grip of a fascination with the origins and evolution of humankind—it had been, after all, a little more than a half century since Charles Darwin's thesis *On the Origin of Species*. Supporters of Darwin believed that man descended from, or was in some intimate evolutionary way related to, the ape. What was needed to confirm this—a supposition that, if confirmed, would further shake the worlds of science, religion, and philosophy—was the discovery of unimpeachable fossil evidence, a palpable link between man and what were then called the "lower" animals.

On December 18, 1912, Dawson and Woodward stood together before the members of the Geological Society of London to reveal that they had made a profoundly important discovery. The skull found at Piltdown, said Woodward, possessed a cranium that had all the characteristics of a hominid, save for the very back of the head where it had connected with the spinal cord: this was much smaller than normal, and indicated that the volume of brain within the cranium had been about 60 percent that of modern man. The jawbone, on the other hand, was not humanlike at all; in fact, it appeared very little different from that of a young chimpanzee. But then there was the pair of molars: these were surprisingly humanlike, suited to a diet that went well beyond the normal rude eating habits of a chimp.

All in all, declared Woodward forthrightly, this mongrel, this hybrid, was indeed the long-sought missing link, the fossil connection between ape and man. To the stunned geologists who were listening in rapt attention, Woodward proposed that this creature, small and shambling, and apparently having lived between 750,000 and 950,000 years ago, should be given a new species name, *Eoanthropus dawsonii*, both in honor of its discoverer and as declaration that the fossil, when living, was

a representative of the dawn of mankind. Not only was Piltdown man firm evidence of this glorious dawn, however, he was also, his presenters implied (to the greater glory of the British empire), a Sussex-born Englishman.

For forty long years this extraordinary theory held sway. There were skeptics, of course, but most particularly in Britain, where paleontologists of the time seemed to be gripped by a visceral wish both to see one of their own venerated and to have one of their countrymen be the link itself, the finding of this creature swiftly became annealed firmly into the evolutionary science of the day—doing, as it happens, considerable intellectual damage to the discipline.

Dawson himself died in 1916, honored as the greatest fossil finder and antiquarian in the country: a memorial was placed at the site of his discovery by a group of the greatest scientists in the land.

And then, in 1953, the rumblings of skepticism burst out into the open.

It had become widely known that buried bones accumulate the chemical fluorine. If these particular bones, the cranium and the half-mandible and the teeth, had in fact lain in the Sussex gravel for three-quarters of a million years, they should have accumulated a very great deal of the chemical. But apparently they hadn't. They exhibited only minimal amounts. Put starkly, this simply didn't compute. Chemistry does not lie, but someone had.

The Piltdown cranium was probably quite old—maybe 50,000 years old. It was of a type found in many archaeological sites in Europe, and thus unexceptional. The jawbone, by contrast, was hardly old at all; moreover, it didn't belong to either the cranium or the teeth. It was the jawbone of an orangutan who had lived for perhaps ten years in Sarawak, on the other side of the world. It had been brought to this site in Sussex, stained with a mixture of potassium bichromate and salt to make it look old, and buried beside the cranium.

Much the same adventure had befallen the pair of teeth, which had been worn down artificially, it turned out, with a metal file. The curious confection that all of these fragments had become was thus not really

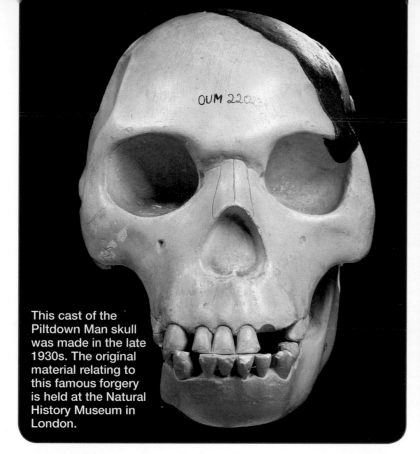

This cast of the Piltdown Man skull was made in the late 1930s. The original material relating to this famous forgery is held at the Natural History Museum in London.

old at all—not one of the bones had anything to do with the Pleistocene, nor anything to do with the others. Nor were they, most crucially, in any way a link to anything of scientific importance.

Piltdown Man, in other words, was a monstrous hoax—a not very clever forgery that nevertheless had fooled most of the British scientific establishment for four decades, decades during which the scientists might have been investigating, with a greater sense of urgency, the real path of human evolution.

As to who perpetrated the hoax, books and essays in vast number claimed this or that figure—among the suspects named were Sir Arthur Conan Doyle (the creator of Sherlock Holmes), the Reverend Teilhard de Chardin (the discoverer of the much more paleontologically respectable Peking Man), and a slew of other eminent scientists. Arthur Smith Woodward, alone, was generally excused from the gallery of potential rogues: this serious and unsmiling martinet (who in his forty-two years with the British Museum took only half a day off for sickness, and that was when he broke his arm) was seen merely as a hapless dupe,

one of the primary victims of this immense forgery (the other being the reputation of British science).

Charles Dawson is now, however, the man around whom the noose of historical judgment has most firmly closed. Subsequent examination of his legacy—his library, his collections—shows him to have been a habitual plagiarist, a habitual forger, the Jason Blair of the geological world. He had told his friend Arthur Conan Doyle back in 1909 that he wanted to make the "big find" that would win him a fellowship of the Royal Society, perhaps even a knighthood. In the boxes of relics that he claimed to have found, no fewer than thirty-eight are now known to have been total fakes. Teeth of an ancient mammal had been filed down. An elephant-bone tool had been carved with a modern iron knife. A Chinese bronze vase was found to be neither Chinese nor bronze. All of these, and thirty-five more, were faked—dry runs, as it were, for the big find of 1912.

When Dawson died in 1916, the finds died with him: no further missing links have been uncovered in England. The sad bony remains of Charles Dawson's melancholy display of misguided ambition, still named Piltdown

Man, were consigned to the back of the museum, a mere curiosity, of no scientific value whatsoever, now merely a relic of one man's hubris—and a painful, cautionary, warning of the startling credulity into which even the most distinguished scientific community can slip.

Evolution and the Human Skull

IT CAN FAIRLY BE said that in the history of biological science never has so much been imagined by so many on the evidence of so little than among those who have studied the skull and wondered about human evolution.

Theories relating to the abilities of entire populations of human ancestors have been surmised from a single fragment of a cranium . Entire academic careers have been built on a few shards of bone dug from pits in remote African deserts. Reputations have been destroyed by a jawbone. Family dynasties—the Leakeys most famously—have dominated a world circumscribed by just a few ounces of skeletal material.

All of this stems, in part, from two quite simple facts. First, the visible changes to the skull that mark the passing of the time that humankind has existed on earth, are terribly subtle; and second, the discovery of fragments of bone displaying these changes are terribly infrequent. The subtlety of the changes—the reason so much store is set by each fragment that is found—follows from another simple fact: that the evolutionary processes that have led to the present-day human skull, the part of the human skeleton that most clearly demonstrates evolutionary change, have taken place in what, in biological terms, has been the blink of an eye.

If we accept that the first cellular life appeared on earth two billion years ago, the first chordate half a billion years ago, early mammals 200 million years ago, the first apes just 28 million years ago, the great apes 13 million years later, and apes that walked upright-ish through the valleys of Ethiopia 10 million years later still—we realize that creatures looking more or less like humans,

behaving more or less like humans, with heads and skulls increasingly human-like, and which in the end evolved into humans, have been around for, at the very most, a total of 3 million years, a tiny fraction of the history of our earth.

One can, of course, go back many millions of years before this, and come up with all manner of creatures that are on the particular evolutionary conveyer belt that ended with the speciation of mankind. Many are extinct, though some are not: when, forty million years ago, the early primates diverged into two basic kinds, the wet-nosed and the dry-nosed, we see many animals that are represented in the skulls collection here, on the one hand the first lemurs and lorises (which are wet-nosed), and on the other, their dry-nosed brethren that include early tarsiers, monkeys, and, most importantly for the human story, apes.

We can readily see the differences between the skulls of monkeys and apes. But once the monkeys have broken away, some twenty-eight million years ago, it then

becomes a little trickier, asking quite a lot of the untutored eye to discriminate within the world of true apes, greater or lesser, to see differences between the skulls of gibbons, gorillas, orangutans, chimpanzees, and bonobos.

And it becomes still more difficult to assess the structural changes of more recent times, once the human line and the chimpanzee line fully separated.

The date of this separation, the date of what is known as the LCA, the Last Common Ancestor of apes and man, is now put at between five and seven million years ago, a time notable for a marked and momentous shift in the structure of the skeleton: the arrival of a creature whose spinal cord enters the cranium not from the side, but from below.

For this is the moment when an ape first stood up—when one group of primates hauled themselves up from walking on all fours, from living in trees—and instead started walking, upright, on the ground, on two legs. This advanced kind of ape was

 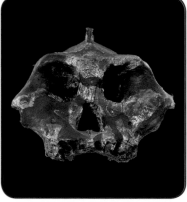  

Australopithecus africanus

A cast of an *Australopithecus africanus* specimen discovered by Robert Broom in 1947. Named "Mrs. Ples," it was one of the most complete early hominid skulls and the location of the foramen magnum indicates the skull was balanced perpendicular to the spine. This results in an obligate, bipedal stance that can support a far greater weight, leaving the gates open for a rapid evolutionary increase in brain mass.

Paranthropus aethiopicus

A replica of the "black skull," the best specimen of this early hominid species, which was found in western Kenya and has been dated to 2.5 million years ago. It has the smallest brain case of any hominid species yet discovered, a pronounced sagittal crest and enormous zygomatic arches. All of these primitive features allow a strong, grinding bite. The skull bears a striking resemblance to the modern gorilla.

Homo habilis

Homo habilis fossils were found in the 1950s close to the Tanzania-Kenya border. Dating back over two million years, they were characterized by an increase in brain size and the first known use of stone tools in our evolutionary history. Recent evidence puts *Homo habilis* (literally meaning "handy man") as a contemporary of *Homo erectus*. Importantly, fossils have shown that bulges in areas of the brain case indicate an enlarged Broca's area, a part of the brain essential for language. This specimen is a cast.

Homo ergaster

From about 2.5 to 1.7 million years ago this hominid species was migrating out of Africa and across the globe. The skull from which this cast was taken was found in Africa. It is an intermediate between the modern *Homo sapiens* and ancestral hominids such as *Paranthropus aethiopicus*. The braincase approaches the size of modern human's but is shaped differently: it is narrower and lacks the bulbous shape and steep forehead our skulls have. This indicates certain brain structures crucial to higher-level cognition (such as the temporal lobe) were less developed than ours.

bipedal—the first true hominid, the first of the line of creatures that would lead to modern man.

The evolution and advancement of this line of creatures—and the consequent changing shape and appearance of their heads and skulls—has been exceptionally swift. In little more than one-tenth of 1 percent of the time that life has been in existence on our planet, the head of an *Australopithecine*—one of the earliest post-separation bipedal hominids, and a presumed direct ancestor of man—has been progressively reshaped into a line of heads that has now given us Albert Einstein, Malcolm X, Mao Zedong, Sitting Bull, and Mahatma Gandhi, among billions of other examples of *Homo sapiens* of both today and the recent, retrievable past.

Three million years is a very short time indeed for this to have been accomplished—the principal reason there seems at first sight to be only a relatively small amount of evolutionary change to the structure of the human skull.

If we look closely, we see that the angular relationships of the bones are different, as is the shape of the jaw and the size of the braincase and the height of the forehead and the protuberance of the brow. Still, very basically, a skull from three million years ago looks remarkably the same as one from today.

There is no argument, of course, that the brain inside the skull has evolved during that time in the most profound manner—enlarging the abilities of early man, who grunted and could barely shape rock into tools, into those of a creature that can compose requiem masses and devise atomic reactors, divine the sequence of the genome, and contemplate his own origins (not to mention his amazing ability to contemplate). Superficially, though, while between a whale and a dog there is a significant difference in appearance, the difference between an australopithecine hominid and a modern-day *Homo sapiens*, is—at least visibly—not so much.

The most famous australopithecine is Lucy, found in Ethiopia in 1974 (and named Lucy after the Beatles' hit "Lucy in the Sky with Diamonds"). It was to be nearly twenty years before the first complete skull was found; up to that point all discussion of the appearance of an australopithecine skull had been based on a few fragments. Now it was possible to see what the earliest hominids looked like: a brain size of about 400 cubic centimeters, ridges above the eyes that in essence were sited where the human forehead is now, a face that projected outward above the jaw, canine and molar teeth smaller than a chimp's but larger than those of modern humans.

Australopithecines were probably hairy; fossil footprints indicate they moved about on two feet; they had hands—and feet—that could grasp things (as human feet still can, if trained to do so in early childhood); and they almost certainly used simple tools.

A possible descendant of *A. afarensis*—to give Lucy her full species name—is a creature that, if it is a true species, is the last of the pre-*Homo* hominids, *Kenyanthropus platyops*. There is a skull, found by Lake Turkana in Kenya in 1999, but it's in very poor condition, its countless cracks filled with solidified clays that make it difficult to describe. But one thing is clear: unlike Lucy, *Kenyanthropus* has a very much flatter face. It has a small brain and high cheekbones—but, under its nose bone, a plane that is flat rather than projecting, a change that makes it look well on the way to becoming properly human.

And then everything changed.

Two and a half million years ago, *Homo* first appears—a genus defined principally by having a suddenly and irreversibly expanded brain capacity. *H. habilis*, one of the earliest *Homo* species, had half as much brain again as Lucy and her immediate followers; and *H. ergaster, H. erectus* and *H. heidelbergensis*, successors to the wily and inventive habilis, possessed double this amount again—a rate of acceleration of

Homo erectus
The skull that this cast was taken from is very similar to modern human's; only the prominent brow ridges distinguish it. *Homo erectus* stuck around in the fossil record for over a million years, only dying out around 100,000 years ago. There is a debate as to whether this species was a direct human ancestor or in fact a sister species to *Homo ergastor*. What we do know is this extremely successful hunter-gatherer hominid migrated out of Africa and colonized the world.

Homo neanderthalensis
Once portrayed as a brutish caveman, *Homo neanderthalensis* was a very similar creature to *Homo sapiens*, but generally smaller and of a stockier build. The brain case is slightly narrower than that of the *Homo sapiens* so there was less space for advanced structures such as the frontal lobe. This specimen is a cast.

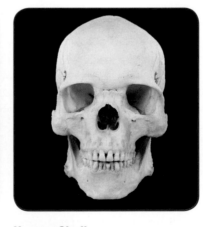

Human Skull
A great deal has been surmised from the human skull in the past with speculation on race, intelligence, and even emotional state being offered.

cranial capacity, and thus presumably brain power, that marks out why modern humans, the only extant survivors of the *Homo* genus, are so systemically different from any creature that had come before, within, or beyond the clade in which they are classed.

The braincases of the skulls of these steadily less and less primitive hominids—there are thirteen of them known so far, not all proven to be or agreed to be distinct species, but all of them extinct—become ever larger and larger as the genus persisted. The face flattens progressively too, changing (to our eyes) steadily from chimplike to humanlike over the years.

In summary: the past three million years have seen the humanoid and early human skull alter, subtly, but importantly; it's enlarged to accommodate more brain, it has lengthened and flattened to allow the steadily diminishing jaws to cope with differing diets; the patterns and numbers of teeth, the size of the eyes, the boundaries of the auditory apparatus, and the relations of face to skull, such that the face becomes smaller and more refined, all have changed, subtly but definitely—slowly revealing creatures with a more sympathetic, more human countenance, no longer beasts appearing to be designed primarily for confrontation. Today's human head, one might say, is more given to smiling than snarling.

Is the human skull still refining and defining itself today—is it still evolving?

It's been tempting to think so. We look, for instance, at the withering into extinction of our wisdom teeth, and we assume this phenomenon to be somehow a consequence of the shrinking back of the jaw, an indication that yes, evolution is still under way.

But the science says otherwise: there are good, nonevolutionary reasons for the vanishing of our wisdom teeth, and much additional evidence to suggest that, in fact, so far as modern humans are concerned anyway—and not simply our skulls, symbolic

though they may be—evolution has all but come to a halt.

Civilization is the principal reason, medicine and technology among the major relevant aspects of civilization. Humankind is increasingly insulated from the very natural selection processes that underpin Darwinian evolution. Put more bluntly, by keeping even the weakest among us alive through natural disasters, attacks of disease, and what once would have been fatal accidents, we may have interrupted the processes by which a species adapts to a less organized and controlled environment. Modern food distribution, disaster relief, and medicine ensure, half-unwittingly maybe, that not only the fittest are able to survive, as once was true, but that everyone human, or a goodly proportion of everyone, now enjoys the opportunity to live long enough to reproduce.

The obvious corollary of that individually happy fact is that, if so many more than previously now live and reproduce, then what we have known as "natural" selection cannot occur, for none of the population is, at least in a genetic sense, so openly exposed to the caprices of nature as all living creatures have been throughout the overwhelming majority of life's presence on earth.

Human equality, in this sense of equal insulation against the worst that nature can throw at us, has truly come about as a result of civilization. The consequence is that the human species may be becoming frozen in a kind of evolutionary stasis.

Is this a good thing, for the population as a whole? To some—to the most pious, perhaps—such an evolutionary standing-still would simply confirm the belief that man, as currently configured, is as advanced, as suited to his surroundings, as "fit" as he is ever likely to be. He is either the child of God or the apotheosis of Darwinian art. Others point out that this stasis may be moot: the very forces that stifle further evolutionary advance of humans also have conferred upon the species the ability to destroy all

macroscopic life on earth.

The debate will continue for a very long time—and all the while the skulls gaze down, their mouths agape, their ears unhearing, their sightless eyes wide shut.

Phrenology

IT WAS INVENTED at the beginning of the nineteenth century by two German physicians named Gall and Spurzheim. It was later abetted by Andrew and George Combe (physician and lawyer, respectively), two of the seventeen children of an Edinburgh brewer. Eventually it attracted millions of followers, spawned societies and clubs (five in London alone), helped spur the creation of immense collections of human skulls, and even initiated the building of a museum devoted to its history, owned by a member of the British parliament. It seems to have been Spurzheim who concocted its pretentious name (from the Greek for "mind" and "knowledge").

But around the middle of Victorian times the "discipline" of phrenology—which was also known variously by the even less attractive terms physiognomy, cranioscopy, and zoonomy—began to be wracked by internal schisms. Soon it became the target of sarcastic and satirical attacks in the press, rapidly lost adherents, and proceeded to wither and die.

Not quite. As late as 1911 its entry in the celebrated (if now increasingly notorious) eleventh edition of the *Encyclopaedia Britannica* suggested that, in some circles at least, it was still being taken seriously. (The six-page entry on "negro" in that same edition confidently informs us—based largely on phrenological observations—that "mentally the negro is inferior to the white.")

Today phrenology is almost universally derided. It's dismissed as pseudoscience. Its legacy is found principally in antique shops: as with many other of history's oddities, it has its modern enthusiasts, who in this case have

> By keeping even the weakest among us alive through natural disasters, attacks of disease, and what once would have been fatal accidents, we may have interrupted the processes by which a species adapts to a less organized and controlled environment.

no interest in its ideas but enjoy looking for the curio skulls that were once found prominently placed in phrenologists' consulting rooms. These skulls are made (if genuine) of bone or (if modern copies) of plastic.

The originals are particularly sought after because they are marked so prettily with delicately etched lines that divide the skull into the areas beneath which lie the locations—or the supposed locations—of a person's moral, spiritual, and intellectual character. These "phrenological skulls" can indeed be most attractive things, but not to be taken at all seriously, none the less prized as amusing, decorative items for the drawing rooms or the offices of the chic.

The five central principles of phrenology begin reasonably enough but soon become progressively more out of touch with reality.

The first holds that the brain is the organ of the mind: no major argument there.

The second states that man's mental powers can be enumerated as a set of identifiable faculties—which is fine, so long as those "faculties" are sensibly and logically described.

The third holds that these faculties are innate, and each can be located in a specific region of the brain—which, up to a point, is true enough so long as the faculties are, as mentioned, sensible and logical.

After that, things get very strange.

The fourth principle of phrenology has it that the size of each region bears a direct relation to the degree to which that particular faculty is expressed in the character of the individual. In other words, if you happen to be a very generous person, then the part of your brain where the faculty of generosity is found will be very much enlarged. It would be tempting to say that such an assertion is the purest poppycock; but let us instead merely note that, during phrenology's brief period of fashion, that claim remained stubbornly unproven.

It is the fifth and final principle that truly takes the cake: it held that the sizes of these purported discrete functional regions of the brain are reflected in the surface and shape of the skull. This principle was the key to the "practice" of phrenology.

The details of your skull would reveal much about you to anyone who was skilled enough in knowing how to examine it: a phrenologist,

A reproduction of a glazed phrenological bust by Lorenzo Niles Fowler (1811-1896) displaying forty-two phrenological organs on one hemisphere and their seven collective groupings on the other.

in other words. If a phrenologist found a bulge in your head above the region where the generosity faculty was located, he could tell you, merely from having felt the shape of your head, that you were a very generous person.

And this was what stirred up all the interest: you allowed in a qualified phrenologist (for only a modest fee) to cause his delicate fingers to wander across your skull, and every aspect of your personality, your moral standing, your proclivities, and your tendencies could be detected, as if they were printed and published in a book.

Susceptible Victorians flocked in droves to the newly built phrenology clinics and there listened to nonsense, won some comfort, were seldom presented with a remedy or a cure, but were always handed a bill—until, that is, the practice was revealed to be no more valid than other quackeries, such as bleeding, trepanning, mesmerism, and the use of the ducking stool.

The flaws in the system, invisible to the gullible, had long been evident to the sensible. And the flaws in the system are many—one of the most basic problems being the notion that enlargement of parts of the brain necessarily leads to shape changes in the inside and then on the outside of the skull. This is simply not true.

The unexamined assumption seems to have been that differential growth inside the skull would cause the outside of the skull to change shape too. But no one seemed to ask how the skull could be affected by growth, or lack thereof, in each region of the brain.

The skull's bony structure is very hard, inside and out. When the tissues of the brain expand — as in a disease such as cerebral meningitis, for instance—the pressures that build up against the inflexible inner wall of the skull cause intense pain. The swelling and consequent compression can damage the brain. But what never happens in meningitis, or in any other severe inflammation of the brain tissue, is damage to the skull itself. It doesn't bend or distort or suddenly give way under the pressure, nor does it acquire bumps or bulges.

In 2006 in a North London hospital six men took part in a drug test that went seriously

wrong. They all suffered the horrific effects of what is known as a cytokine storm—a dreadful allergic reaction to the drug—which included massive swelling of their body tissues, especially in and around the head. One man reported that he looked like the Elephant Man. The tabloid press seized upon his words. "We Saw Human Guinea Pigs Explode" was one headline.

But it turns out that this unfortunate man's skull, quite specifically, did not change its shape: his face had swollen, together with all the tissues of his neck and his shoulders, his eyes and ears and nose. He looked terrifying. But his skull remained as it had been—that was part of the problem, the source of his intense head pain.

Had an honest phrenologist happened to have run his hands over the hard parts of this unhappy man's head, he would have found it essentially unchanged in shape and size. Because what is true of all skulls was true of his: once maturity has been achieved, skulls do not change shape . They can be fractured or be penetrated if sufficient external force is applied, but that is essentially all they ever do.

The other aspect of phrenology that eventually brought it down was the vague and dreamy way that the various "faculties" possessed by the human brain were described. A manual sold in 1876 by one of its best-known practitioners, a man named Professor L.N. Fowler ("who has decided to open Permanent Rooms for Phrenological, Physiological and Medical Consultations at 107 Fleet Street, EC, with office hours from 10am till 5pm"), presented some thirty-nine "faculties," their locations marked on his Patented Map of the Cranium (available both in his consulting rooms and as the frontispiece of his *Self Instruction Manual* published for some ten shillings by Messrs Tweedie of 337 The Strand, London). "Half a million copies of the lectures which describe these faculties have so far been published," Professor Fowler tells us.

And what faculties does the brain possess,

and where are they? Just above the eye, he says, are located those of Color, Order, and Calculation. At the very summit of the skull lie Benevolence, Veneration, and Firmness. Behind the ear, Combativeness, Amativeness (Love Between the Sexes), and Vitaviveness (Love of Life, for the uninitiated). In that place where one hopes the hat remains, at the back of the skull, we see Conjugality and Parental Love (and, presumably, to keep that hat in place, perhaps Adhesiveness?).

Hence: let those skilled fingers play across your head, and prepare to be alarmed or comforted. Let Professor Fowler declare there to be a significant bulge beneath the back of your ear, and you are clearly a Don Juan, or worse. A lump at the very top of your head, and you are a person who shows respect. No swelling at the top side of your eye? You lack the faculty of Calculation (which explains, the phrenologist thinks, why his charge of two guineas for the phrenological consultation will seem like almost nothing at all to you).

It was all pure bunkum. And it died in an odd way.

Given the rash of new words that attended the appearance of this non-science, one can only imagine the irritation in the dictionary-making world. One of phrenology's many foes—one of the Victorian figures who helped hasten its decline as a respectable diagnostic tool, by burying it with ridicule—was no less a figure than Dr. Peter Mark Roget, the man who gave the world *Roget's Thesaurus*.

Roget, himself a physician, derided phrenology, and he helped destroy it. What endures is Roget's lexical antipathy: look up Phrenology in Roget. You will find it grouped there with *palmistry, prophecy,* and *divination.* In other words, it's in the good company of other amiable nonsenses.

Phrenology: harmful and harmless by turn; now just a fancy, a jeu d'esprit—albeit one that has left us a million skull maps and models as its rather elegant, and eminently collectable, legacy.

Fish

Seahorse
Hippocampus sp.

THIS DELICATE SEAHORSE skeleton has not been identified as particular species but is representative of the family. Seahorses have no scales and live cryptic lifestyles hidden among weeds. Famously, the male gives birth to tiny, fully developed seahorses, the female having deposited her eggs in the male's brood pouch for incubation.

Kingdom: Animalia
Phylum: Chordata
Class: Actinopterygii
Order: Syngnathiformes
Family: Syngnathidae
Genus: Hippocampus
Behavior: Carnivore/
Aquatic

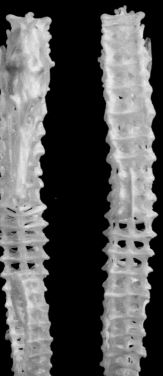
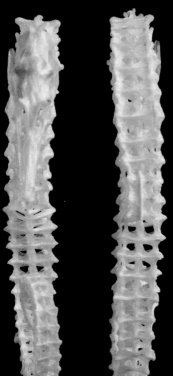

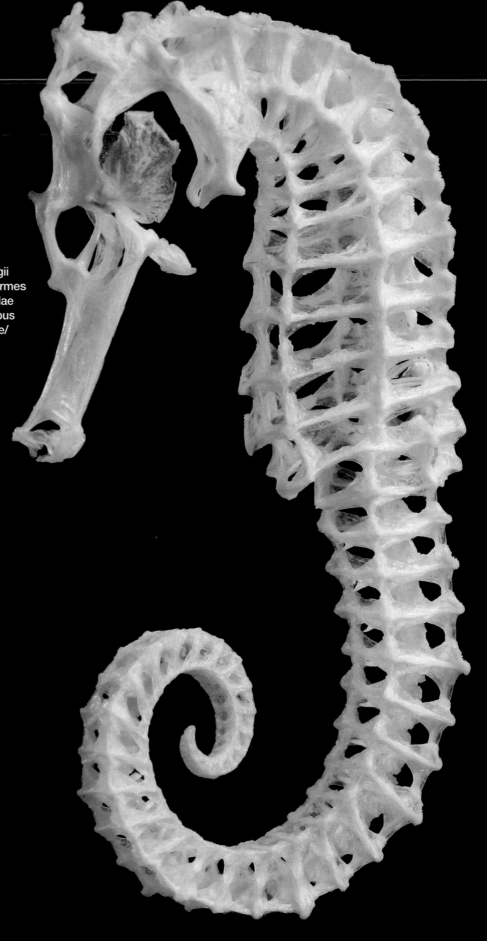

DUDLEY'S NOTES:
This is an absolutely fantastic skeleton prepared by a guy in America. He used domestic beetles in laboratory conditions. The beetles start from eggs and only the smaller beetles are able to get inside the skeleton and eat away at the flesh. As the beetles grow larger, they aren't able to get into all the nooks and crannies. It's quite a skill to be able to use domestic beetles.

European Plaice
Pleuronectes platessa ◁

CHARLES DARWIN WAS BAFFLED by flatfish anatomy. "During early youth," he noted, the eyes "stand opposite to each other. ... Soon the eye proper to the lower side begins to glide slowly round the head to the upper side. ...The chief advantages thus gained seem to be protection from their enemies, and a facility for feeding on the ground." This remarkable skull arrangement arises during the youth of every flatfish, where the symmetrical larval skull undergoes a metamorphosis to produce an asymmetrical juvenile. One eye "migrates" up and over the top of the head before coming to rest in the adult position on the opposite side of the skull.

Kingdom: Animalia
Phylum: Chordata
Class: Actinopterygii
Order: Pleuronectiformes

Family: Pleuronectidae
Genus: Pleuronectes
Behavior: Carnivore/
Aquatic

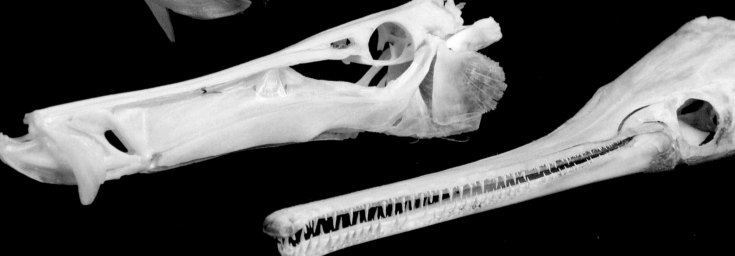

Trumpetfish
Aulostomus maculatus △

THIS EXTRAORDINARILY NARROW and elongated skull belongs to a trumpetfish—a fish that often swims vertically, pretending to be a piece of coral or a weed stem. If prey swims beneath the trumpetfish's mouth, it will suck the morsel up in an action known as pipette feeding.

Kingdom: Animalia
Phylum: Chordata
Class: Actinopterygii
Order:Syngnathiformes

Family: Aulostomidae
Genus: Aulostomus
Behavior: Piscivore/
Aquatic

Longnose Gar
Lepisosteus osseus △

THE LONGNOSE GAR has a splendid set of jaws. Specimens weighing more than twenty-three kilograms (fifty pounds) have been claimed by Texan anglers (who traditionally spear them). They are nocturnal hunters of smaller fish and are themselves hunted—there is one in the jaws of the photo accompanying the American alligator skull.

Needlenose Gar
Kingdom: Animalia
Phylum: Chordata
Class: Actinopterygii
Order: Lepisosteiformes

Family: Lepisosteidae
Genus: Lepisosteus
Behavior: Piscivore/
Aquatic

Barracuda
Sphyraena sp.

AS YOU WOULD EXPECT from a large predatory fish such as a barracuda, this skull has a fine set of teeth and jaws. It also has huge orbits—not all that common in fish skulls. It has an elongated, pike-like head and a distinctive underbite: the lower jaw is slightly longer than the upper. There are twenty-seven known species of barracuda.

AKA: Sennet
Kingdom: Animalia
Phylum: Chordata
Class: Actinopterygii
Order: Perciformes
Family: Sphyraenidae
Genus: Sphyraena
Behavior: Piscivore/Aquatic

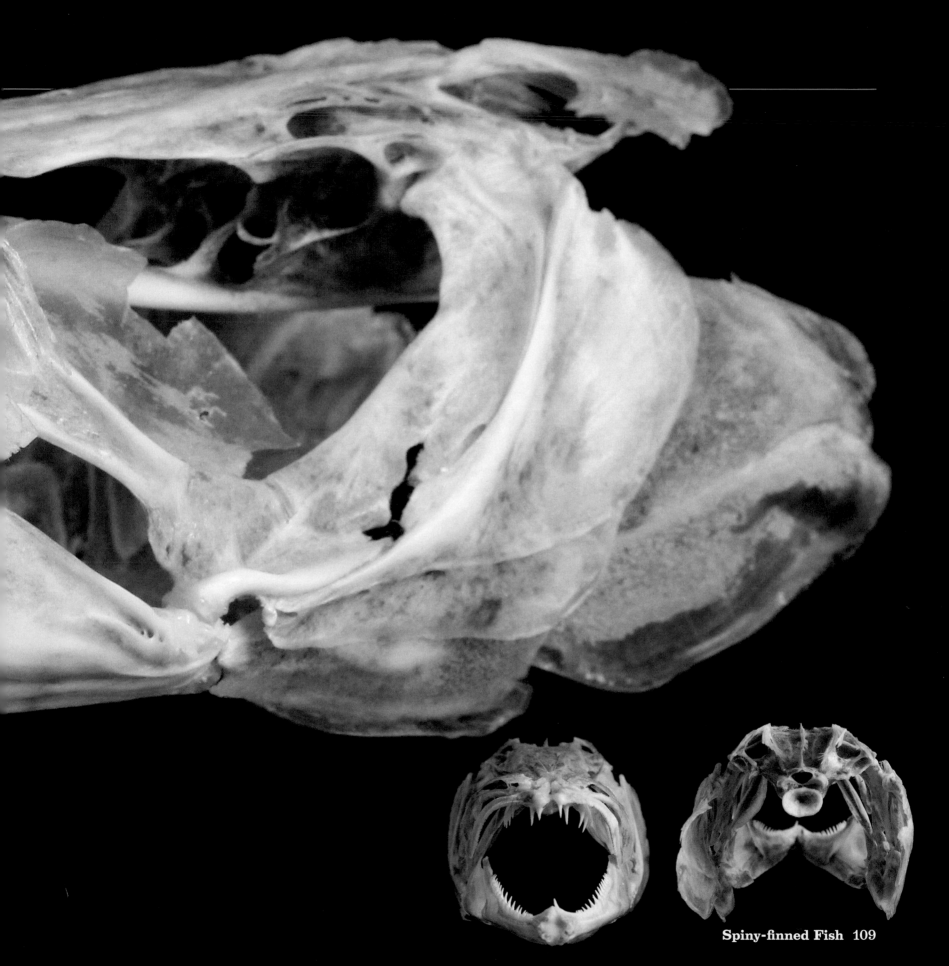

■ Spiny-finned Fish

Black Jack
▷ *Caranx lugubris*

THIS SKULL HAS an impressive, dinosaur-like fin on the top of the head and wonderfully delicate sclerotic rings in the eyes. The black jack is a wide-ranging tropical species, important to local fisheries.

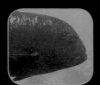

AKA: Black Trevally, Black Kingfish
Kingdom: Animalia
Phylum: Chordata
Class: Actinopterygii
Order: Perciformes
Family: Carangidae
Genus: Caranx
Behavior: Carnivore/Aquatic

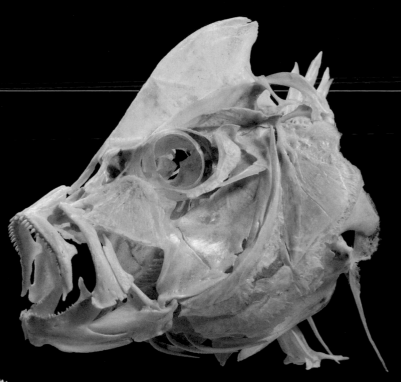

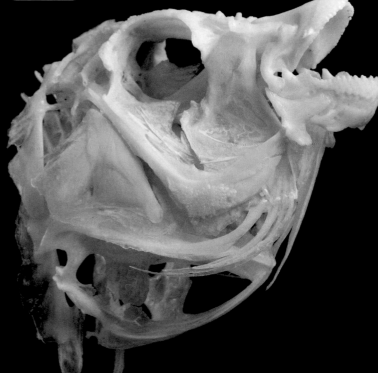

Redtail Parrotfish
◁ *Sparisoma chrysopterum*

THE BEAKLIKE MOUTH of this skull lends the parrotfish its name. The redtail parrotfish uses its closely packed teeth to rasp algae from coral and other rocky surfaces (an important part in generating sand around the reef). Parrotfish teeth grow continuously, as they would otherwise be worn away by grazing.

Kingdom: Animalia
Phylum: Chordata
Class: Actinopterygii
Order: Perciformes
Family: Scaridae
Genus: Sparisoma
Behavior: Herbivore/Aquatic

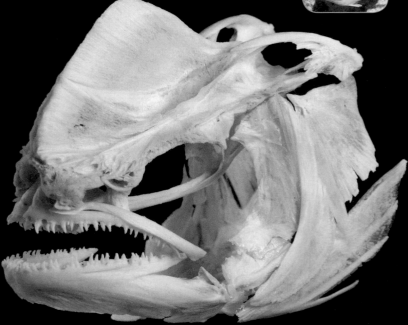

Mahi-Mahi
▷ *Coryphaena hippurus*

THE MATURE MALE mahi-mahi has an impressive sagittal crest. It is a fast-swimming predator with large eyes and recurved teeth adapted for gripping its prey. They are relatively common, and delicious, making them a sustainable choice for supper.

AKA: Common Dolphinfish
Kingdom: Animalia
Phylum: Chordata
Class: Actinopterygii
Order: Perciformes
Family: Coryphaenidae
Genus: Coryphaena
Behavior: Piscivore/Aquatic

Atlantic Wolffish
Anarhichas lupus

THE ATLANTIC WOLFFISH eats hard-shelled mollusks, crabs, and urchins and need the dental apparatus to cope with such well-protected morsels. Within the mouth there are bony structures used for crushing shells and carapaces; around the edge of the mouth are sharp teeth that can grip and tear prey. You can also see a robust bone structure that supports the jaw, allowing the strong muscle attachment necessary for the wolffish to tackle the toughest of marine prey. In Iceland wolffish are known as *steinbítur*—literally translating as "stone biter."

AKA: Seawolf,
Atlantic Catfish
Kingdom: Animalia
Phylum: Chordata
Class: Actinopterygii
Order: Perciformes
Family: Anarhichadidae
Genus: Anarhichas
Behavior: Carnivore/
Aquatic

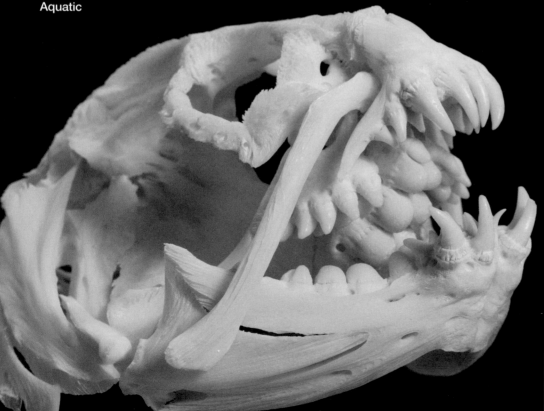

DUDLEY'S NOTES:
I thought monkfish were ugly but the wolffish is even uglier. It's got the most impressive array of teeth that grow in all different directions. Inside the mouth there are crushing molars.

Gray Triggerfish
Balistes capriscus ▷

THIS GRAY TRIGGERFISH has impressive teeth at the end of its jaw. As protection against predators, triggerfish can erect two dorsal spines: the anterior spine is locked in place by the erection of a short second spine. It can be unlocked only by depressing the second, "trigger" spine, for which the fish is named.

Kingdom: Animalia
Phylum: Chordata
Class: Actinopterygii
Order: Tetraodontiformes
Family: Balistidae
Genus: Balistes
Behavior: Carnivore/ Aquatic

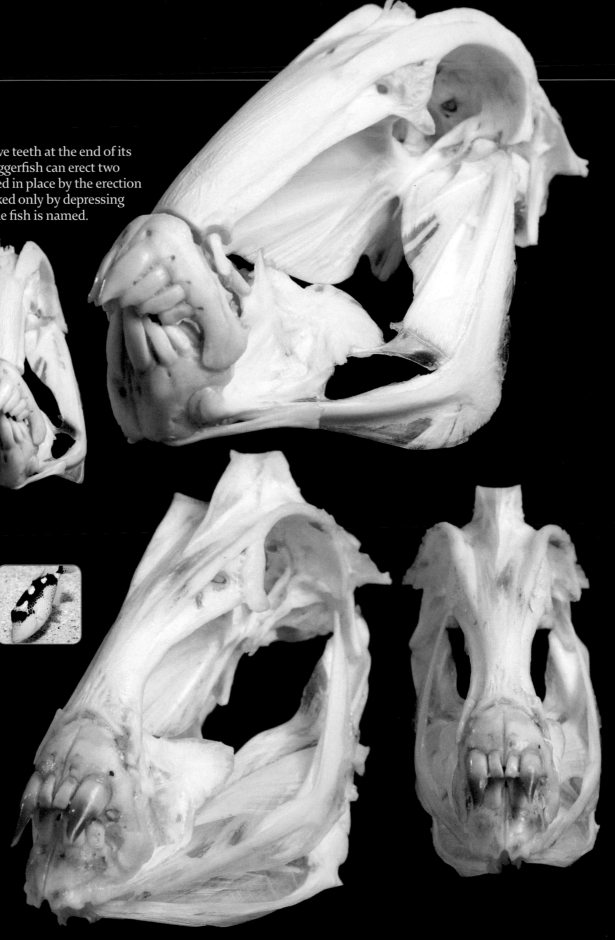

Starry Triggerfish
Abalistes stellatus ▷

THIS SPECIES IS BEST known as the character Dory in the animated film *Finding Nemo*. Triggerfish are brightly colored inhabitants of coral reefs; their mouths are lined with rather vicious teeth to crush small crustaceans and have been known to deliver a nasty bite to scuba divers when they feel threatened. They diverge from their Hollywood depiction by being one of the more intelligent fish on the reef.

AKA: Chicken Fish
Kingdom: Animalia
Phylum: Chordata
Class: Actinopterygii
Order: Tetraodontiformes
Family: Balistidae
Genus: Abalistes
Behavior: Carnivore/ Aqualic

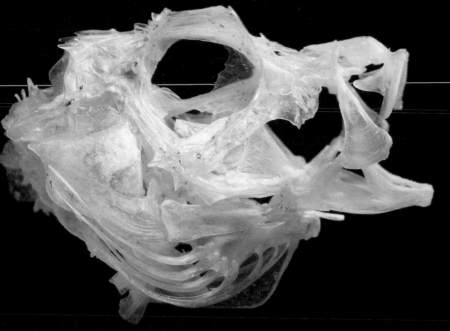

Broadbarred Firefish
Pterois antennata ◁

THIS SKULL LOOKS like something made from layers of paper by an origami enthusiast; it's a wonderful example of delicate skull preparation. The broadbarred firefish is ornately beautiful with venomous spines and tentacles used to attract prey.

AKA: Ragged-finned Firefish
Kingdom: Animalia
Phylum: Chordata
Class: Actinopterygii

Order: Scorpaeniformes
Family: Scorpaenidae
Genus: Pterois
Behavior: Carnivore/ Aquatic

Long-spine Porcupinefish
Diodon holocanthus ▷

THIS SKULL OF THIS fish bears a good likeness to the living creature. The jaws of the long-spine porcupinefish are formed from fused teeth and are an adaptation for its diet of hard-shelled creatures such as small mollusks and crustaceans. The porcupinefish can inflate its body when threatened to render it too large and spiny to bite. Porcupinefish are poisonous and used in Chinese medicine.

AKA: Balloonfish
Kingdom: Animalia
Phylum: Chordata
Class: Actinopterygii
Order: Tetraodontiformes
Family: Diodontidae
Genus: Diodon
Behavior: Carnivore/Nocturnal

Fish
■ Piranha and Catfish

Red-bellied Piranha
Pygocentrus nattereri ▷

EVEN THE MOST AMATEUR of naturalists should be able to hazard a guess as to the owner of these jaws. The distinctive triangular teeth of the red-bellied piranha's lower jaw meet the smaller teeth of the upper jaw in a scissorlike action that enables them to slice the flesh off the bones of dead (and sometimes living) animals in the tributaries of the Amazon.

Kingdom: Animalia
Phylum: Chordata
Class: Actinopterygii
Order: Characiformes
Family: Characidae
Genus: Pygocentrus
Behavior: Carnivore/Aquatic

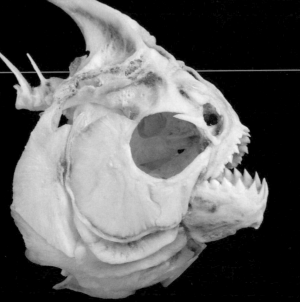

Armored Catfish
Callichthys sp.

ARMORED CATFISH are popular aquarium fish. This complete preserved specimen has only been identified down to the family taxon. It is beautifully prepared, looking rather like a fossil. Catfish are bottom feeders, and the "whiskers" near their mouths are known as barbels.

AKA: Callichthys
Kingdom: Animalia
Phylum: Chordata
Class: Actinopterygii
Order: Siluriformes
Family: Callichthyidae
Behavior: Omnivore/
Aquatic

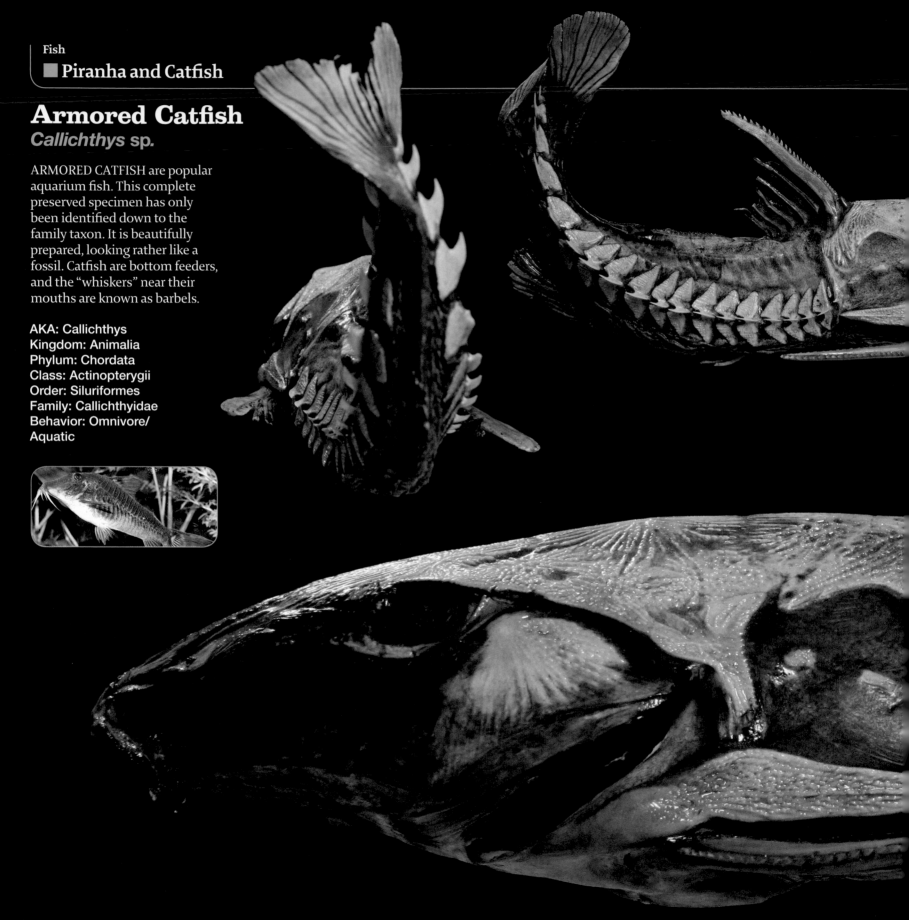

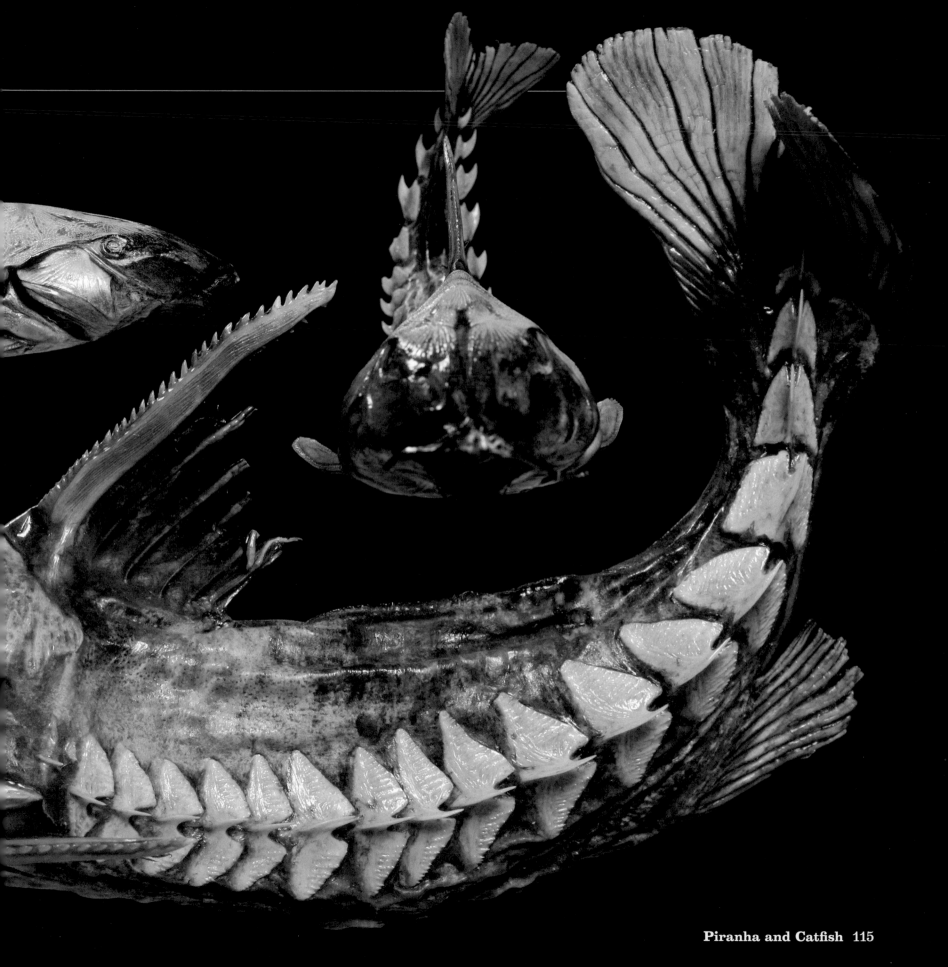

Giant Moray
Gymnothorax sp.

YOU MAY BE FORGIVEN for thinking this was the skull of a snake—and you wouldn't be too far wrong. It's a moray eel. This particular skull has not been pinned down to a specific species. These eels can grow to three meters (ten feet) and have been reported to attack scuba divers. This skull preparation doesn't show the set of pharyngeal jaws that would lie deep in the throat of the living specimen.

AKA: Giant Moray Eel
Kingdom: Animalia
Phylum: Chordata
Class: Actinopterygii
Behavior: Carnivore/ Aquatic

Conger Eel
Conger conger

SINCE THIS SKULL originated in British waters, it is in all likelihood *Conger conger*, the European conger eel. Unlike the tall, narrow skull of the moray eel, this one is flattened dorso-ventrally with several rows of very fine, sharp teeth. Congers are often found in deep waters and like moray eels, are familiar inhabitants of shipwrecks visited by scuba divers. These fish can grow exceptionally large, reaching over 100 kilograms (220 pounds).

Kingdom: Animalia
Phylum: Chordata
Class: Actinopterygii
Order: Anguilliformes
Family: Anguillidae
Genus: Anguilla
Behavior: Piscivore/Aquatic

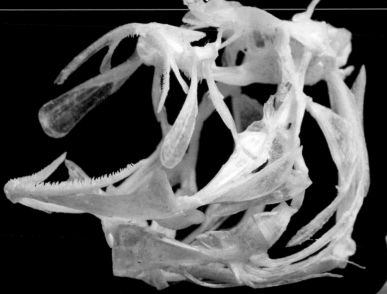

Atlantic Pollock
Pollachius pollachius ▷

POLLOCK IS THE POOR relation to the allegedly tastier (but scarcer due to overfishing) cod. The fish are very similar, although the pollock's lower jaw tends to protrude further than that of the cod.

AKA: Coley, Saithe
Kingdom: Animalia
Phylum: Chordata
Class: Actinopterygii
Order: Gadiformes
Family: Gadidae
Genus: Pollachius
Behavior: Piscivore/Aquatic

Warty Frogfish
Antennarius maculatus ◁

THIS IS A MOST BEAUTIFUL and delicate skull. In the flesh, frogfishes take on all sorts of weird and wonderful disguises, from clumps of seaweed to fronds of coral. They are closely related to angler fish and have a similar ambush approach to their feeding strategy.

Kingdom: Animalia
Phylum: Chordata
Class: Actinopterygii
Order: Lophiiformes

Family: Antennariidae
Genus: Antennarius
Behavior: Piscivore/Aquatic

Common Ling
Molva molva ◁

THIS IS ANOTHER FISH, like the pollock, that your local fishmonger may pass off as cod. Ling is a large deepwater fish possessing a broad, flattened skull with needle-like teeth for gripping its prey (usually other benthic fish and crustaceans).

AKA: Ling Cod, Sea Pike
Kingdom: Animalia
Phylum: Chordata
Class: Actinopterygii

Order: Gadiformes
Family: Lotidae
Genus: Molva
Behavior: Carnivore/Aquatic

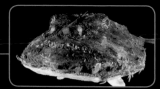

Monkfish
Lophius piscatorius

MONKFISH TYPICALLY have three long filaments sprouting from the middle of their heads. Beautifully preserved in this skull, these are the modified first three spines of the anterior dorsal fin. The longest filament is moveable in all directions and is used as a lure to attract prey. When close enough, the monkfish then seizes its prey with its enormous jaws, devouring them whole.

Kingdom: Animalia
Phylum: Chordata
Class: Actinopterygii
Order: Lophiiformes
Family: Lophiidae
Genus: Lophius
Behavior: Carnivore/
Aquatic

Houndfish
Tylosurus crocodilus

THIS FIERCE-LOOKING houndfish skull could be mistaken for that of a savage stork with teeth! Houndfish can grow to large sizes and are prized by game fishermen not only for the fight they put up but also for the taste of their flesh. They have a reputation for jumping out of the water when attracted to a boat's lights.

AKA: Crocodile Needlefish
Kingdom: Animalia
Phylum: Chordata
Class: Actinopterygii
Order: Beloniformes
Family: Belonidae
Genus: Tylosurus
Behavior: Piscivore/Aquatic

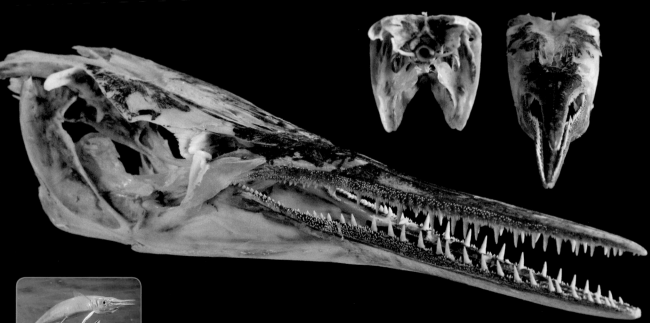

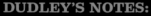

Pike
Esox lucius

THIS PIKE SKULL is photographed in two rotational planes. In the vertical rotation you can see the masses of sharp backward-pointing teeth on the roof of the mouth. Long needle-sharp teeth on the lower jaw also attest to the pike's predatory nature. This skull is not from a particularly big fish, but pike can grow very large indeed; the largest specimens are females.

AKA: Northern Pike
Kingdom: Animalia
Phylum: Chordata
Class: Actinopterygii

Order: Esociformes
Family: Esocidae
Genus: Esox
Behavior: Piscivore/Aquatic

DUDLEY'S NOTES:
This is the first pike I ever prepared and it was a nightmare. I'd gone to a local country park and asked that if they ever came across a pike they should save it for me. When I finally got the pike the ranger neglected to mention that he'd kept it in water for more than a week before phoning me. It was a massive piece of slime and it smelled to high heaven. It nearly cost me my marriage! I ended up with a fantastic specimen though, and even managed to retain all of its teeth.

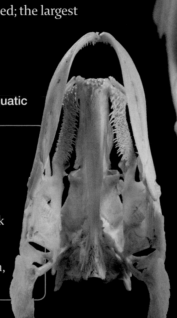

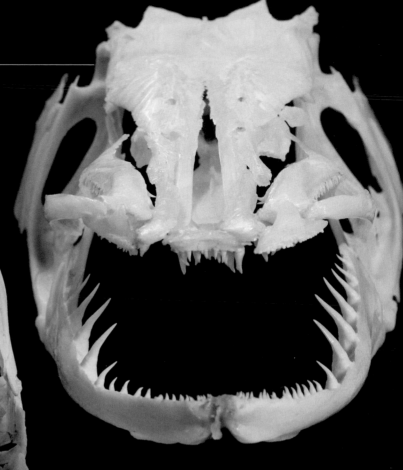

Bowfin
Amia calva

BOWFINS ARE A PRIMITIVE fish in the evolutionary tree. They are related to gar and sturgeon and are considered to be "living fossils" since very similar skulls are found alongside dinosaur remains from as far back as the Jurassic period.

Kingdom: Animalia
Phylum: Chordata
Class: Actinopterygii
Order: Amiiformes
Family: Amiidae
Genus: Amia
Behavior: Piscivore/Aquatic

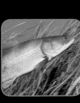

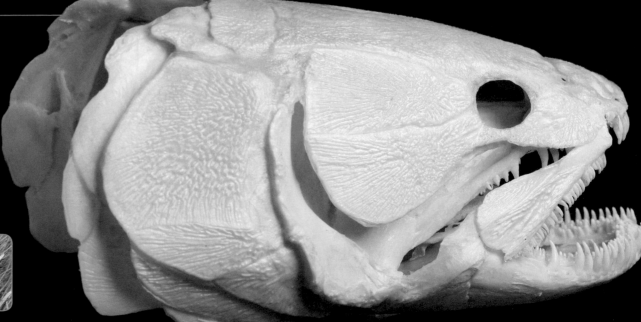

The Iconography of Skulls

THERE ARE 206 BONES in the human body, one more than in a horse, a hundred or so fewer than in the average mouse or dog, and roughly half as many as in the skeleton of the long-extinct stegosaurus. Yet of all of the bony structures found within the animal kingdom—an almost unimaginable proliferation of shapes, varying enormously in length, weight, and function—one structure reigns supreme in matters of mystery, symbolism, magic, and art—and has reigned supreme throughout human history in the biological omnium-gatherum.

And that one is the skull.

The skull—or a skull-without-a-lower-jaw, known as the cranium—is the one part of the animal body, most especially of the human body, that exerts a simply immeasurable degree of power and mystery. The skull alone demands a sense of awe, respect, and fear. It is the one "bone" (actually, an amalgamation of smaller bones, usually about 22 of them) that has become, by dint of playing so profoundly important a symbolic role in human society, a true and lasting icon.

And yet, though the answer would appear at first to be obvious, there has to be the single question: why?

Why the skull—why is it the icon? Why not the pelvis, say—the bone that is the gateway for the newborn? Why not the femur—the skeleton's most majestically large bone (in humans, anyway)? Or, considering the importance of the opposable thumb, why is the iconic structure representing humankind not that of that of the complicated wonder, the bones of the human hand?

We may think the answer obvious, so familiar has the skull become to us all. The eyeless sockets, the firm jaw, the bulging braincase, the teeth in the rictus of an eternal grin—the skull is face and figure, it is man, staring sightlessly back at all who view it. Surely that is enough to make the choice abundantly clear.

But in fact the story of that choice is a little more subtle—its seeming obviousness deserves some explanation. It turns out that the story most probably begins, several score of millennia ago, with cannibalism.

It is generally surmised—though not universally—that early man, in his all-too-frequent times of hardship and famine, practiced cannibalism. The precise details of the anthropophagy, which are unpleasant both to recount and to read, need not concern us here—except those giving evidence that ancient man seems to have accorded varying significance to the parts of the body he decided to consume. Most students of the ancients agree that, when cannibalism occurred, the diners most probably ate thighs and buttocks and forearms with relish and abandon; but when it came to eating heads, or the contents of heads, they appear to have done so in a much different manner, one that treated the head structure and brain container—the skull, in other words—with an unusual reverence: with ritual, and with the frequent application of magic.

This is the best explanation for the disproportionately large number of skulls and skull parts that are to be found today in human or hominid archaeological sites. Had all the bones of the dead been treated alike by the remaining living—and assuming that all bones decay at a similar rate, which science shows that they do—then we would surely expect to find as many spinal columns in gravesites as skulls, twice as many femurs and tibiae and humeri as skulls, ten times as many finger- and toe-bones, and so on.

But that's not the case: ancient graves and burial grounds are the repositories for many more skulls than mere statistics suggest there should be. Moreover, there is something else: many of the skulls are cut or marked or in other ways incised—most cases involving especially the large hole where the spinal cord emerges at the back of the skull, which is typically damaged and enlarged in some way.

All of these factors—the oddly large numbers of skulls, and the cuts and the marks made to them—tend to support the notion that ancient man regarded the skull with unusual esteem, that he often retained a skeleton's skull after death, and that he treated it in a manner very different from the way that he treated a leg bone, say, or a rib or finger.

The marks and cuts around the back-of-the-skull occipital bone and the nearby foramen magnum, as the hole behind it is known, allow one to speculate further, especially since some present-day rituals observed by anthropologists in a small number of remote communities around the world result in the same skull mutilations. It appears that from time immemorial, and in some cultures still today, particular regard has been paid to the human brain—and that by cannibalistically consuming it, as distinct from consuming the flesh from elsewhere on the body, the consumer expects somehow to absorb the wisdom, the knowledge, the personality, even the identity, of the previous owner.

The brain, in other words, has long been honored—in fact long before humankind had any knowledge of what the brain really did. There was some kind of ancient hunch: that the gray matter secreted above and behind the face and beneath the hair played some unknown but uniquely important role in creating and sustaining the essence of its owner. This gray matter was mysterious; it was marvelous.

The corollary to all of this was then plain to see: if the brain is precious and special, then its bony encasement, its protective sheath, is to be accorded a special reverence. It is, after all, the part of the body chosen—by God, by Fate, by evolutionary adaptation—to be the guardian of the delicate stuff inside that invests the human not only with life but, at least in some cases, with wisdom.

So it is hardly surprising that magic—a kind of harnessing of the secret forces of nature, a time-honored wedding of man and mysterious mechanics—became in short order a component of rituals involving the skull . Since the reverence paid to this confection of bones was in itself stubbornly inexplicable, and since for millennia the function of the gelatinous mass of brain inside also remained stubbornly unknown, it fell to magic and magicians to explain its supposed powers.

But ancient man knew these powers were only temporary. The brain itself would eventually vanish—it would decay, or more likely would be eaten by hungry animals, including humans, and lustily so. Yet while the brain went, the skull did not. There it remained, solid and white, apparently imperturbable and eternal—with the result that skulls were preserved loyally, as talismans, as objects of worship, as items of remembrance, as subjects of veneration, and thus the skull became an everlasting icon.

Most animals' faces are entwined with musculature that allows them to form and reform themselves into different expressions—they can express anger, surprise, pleasure, pain, defiance, disgust. The more complex the brain within, the more numerous and subtly different the faces that can be made: a human brain can be coupled to a face to launch a thousand shapes, maybe even more. But once the owner dies, and the flesh and fat and muscle and nerve fiber is undraped from it, the skull is left frozen in an aloof expressionlessness. No longer a face, the skull becomes ghost-like, impenetrable, mute, and terrifying, still recognizably human but no longer alive. And recognizably durable, as the brain and the rest were not. This, too, is why the skull has become an icon—for not only does it represent the soul of man, the essence of its kind—but it is also freighted with the fearful reminder that this is what we will all one day be reduced to. It reminds, it hints, it nudges at the conscience, it warns.

And as such, it is everywhere: the bones we call the skull, of the 206 bones in the human body, fashioned by circumstance into the most universally recognizable of symbols, a lifeless representation of the human antithesis, an icon both fascinating and chilling at the same time.

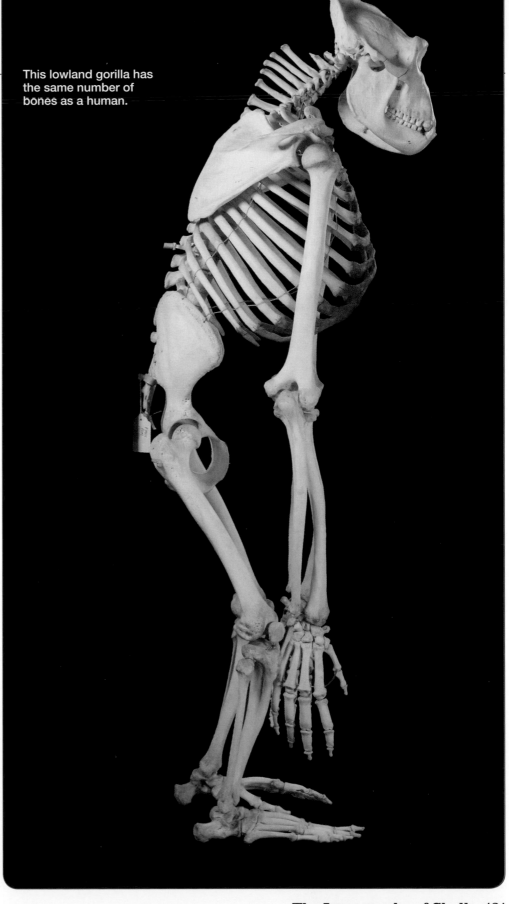

This lowland gorilla has the same number of bones as a human.

The Skull as an Icon of Death

SCANNING THE PAPERBACK shelves in my library, one fat volume in particular stands out. It's black, with the title and author's name written in a strangely corroded Gothic typeface. But the feature that makes the book most noticeable appears dead center on its spine: a human skull, eye sockets exceptionally large, nasal cavity deep and dark, its teeth, all intact, fixed in a ghastly and permanent grin.

The author is Edgar Allan Poe and the volume is *The Complete Works*—and they are indeed all there, the stories, the essays, the poems, just about all of them concerned with the grisly and the grotesque, the dying and the dead, killers and the killed. "The Fall of the House of Usher" is there; so is "The Murders in the Rue Morgue," along with "The Pit and the Pendulum," "The Oblong Box," "The Masque of the Red Death." And all of these grimly classic contents have been summed up by the book's designers with the symbol that, above all others, we associate with mortality, death, decay. See it anywhere—on a book jacket or a gravestone—with its gaping sockets, off-white bone, grinning teeth, and dome of hairless, skinless, bloodless curvature, and you know in an instant: whatever is within or below will be uncomfortable, unpleasant, more than likely concerned with the end of life, and, more often than not, enigmatic, tragic, and violent.

But this has not always been so. In Asia, in Egypt, and in classical Mediterranean cultures, there was no suggestion that any part of the human skeleton—least of all the skull—represented death. Hypnos and Thanatos, the mythological deities of sleep and death, respectively, were regarded in the *Iliad* as twins; those who died inhabited the Land of the Shades as pale ghosts, beings-that-were-not-beings, lacking substantial form or shape and thus requiring no recognizable structure, bony or otherwise, to support their ephemeral dream state as they haunted the world.

However, come the two centuries of the Hellenistic period—the time in Greek history that followed the death in 323 BC of Alexander the Great, a time of swift and dramatic spread of Greek culture through Europe and Asia Minor—this view of the afterlife experiences a transmutation. The hard endoskeleton around which human flesh is wrapped comes forth into prominence. The ghost is made solid, and the skeleton, especially the skull, begins steadily to make its appearance on gravestones and in art as a symbol of the death that had befallen its owner.

In Boscoreale, the villa on a hill two miles north of Pompeii, excavations at the beginning of the last century turned up, among the frescoes and the jewelery, a beautiful silver beaker, adorned with raised images of skeletons. Lord Rothschild acquired the treasures and presented them to the Louvre. A close examination will reveal on the famous beaker a seminal example of the time's swift change in the human perception of bones: there's one skeleton here shown holding a cranium, and beneath it the words: This is Man. For the first time, a human being in death is co-equal with a skull—a watershed moment. The skull had begun its steady rise—at least in Western culture—to become the ultimate and iconic signifier of death.

It was certainly such, three centuries later, when the Greek satirist Lucian wrote his *Dialogues of the Dead*. In perhaps the best known of these satires, Hermes is giving a tour of Hades to a newcomer, Menippus, and at a crucially dramatic moment presents him with a pile of bones and skulls, pointing to one and telling his visitor, with a flourish, "This skull is Helen." "What?" cries the incredulous Menippus. "This skull is of the woman whose face launched the thousand ships?" Oh yes, replied Hermes, for "in the hour of their bloom these unlovely things were things of beauty. ..."

Thus did the skull steadily rise to become the premier symbol of the dead—maybe

As well as a prominent skull, this Roman memento mori mosaic from Pompeii (30BC-14AD) also features a plumb line, suggesting death as the great leveler.

Hamlet and Horatio in the Cemetery (left), Eugène Delacroix, 1839.

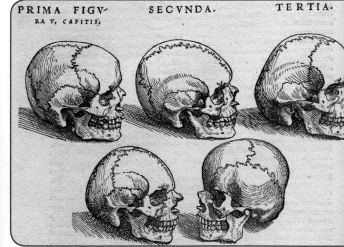

Skulls (*above*) from Vesalius' *De humani corporis fabrica* (1543).

appearing a little less frequently in the art of the Middle Ages, which seems to have fewer death's-heads and skeletons than in Hellenistic times, but then returning to the scene with a vengeance by the thirteenth and fourteenth centuries.

Why the interregnum, why the return? Many scholars have suggested that the Black Death and kindred pandemics present the key to the consolidation of skull-as-death-symbol: it has much to do, they say, with the fact that, between 1348 and 1350, in cities and villages from London to Dubrovnik, from Messina to Stavanger, terrified citizens became all-too-familiar with—almost inured to—flopping, fleshless heads slumped from the ends of a thousand charnel wagons.

In any case, by the sixteenth century there was no doubt: Shakespeare, for instance, gave us the most famous of all skulls: that of Yorick, in *Hamlet*.

William Shakespeare was familiar, no doubt, with the extraordinary drawings of the Brussels-born anatomist Andries Van Wesel— Vesalius, as he is better known. It was Vesalius who, for instance, first drew scientifically accurate images of the human body, of its musculature, of its circulatory system — and of its skeleton. Vesalius's image of a skeleton contemplating its own cranium was well-enough known in sixteenth-century London for an educated figure like Shakespeare to have become aware of it. Was this the model for Hamlet's contemplation of Yorick's skull? Or was the playwright aware of Lucian, and his presentation to Menippus of the skull of Helen? Literary scholars have argued the point for years, but for this account one thing is made abundantly clear: that, with these famous lines from *Hamlet*...

Alas, poor Yorick! I knew him, Horatio;
a fellow of infinite jest, of most excellent fancy;
he hath borne me on his back a thousand times;
and now, how abhorred in my imagination it is!
My gorge rises at it. Here hung those lips that I have
kissed I know not how oft.

...now the idea of the human skull as the most dramatic reminders of mortality, the skull as memento mori, was firmly annealed into the public consciousness.

Imagery of the young and the beautiful holding or caressing or contemplating the skull—the pink and warm connecting with the gray and cold, the living in intimate association with their very antithesis—is, all of a sudden, everywhere: the Dance of Death, cadaver tombs, the Grim Reaper, the works of Franz Hals, of van Leyden, Delacroix, Hans Memling, Fra Angelico, Poussain, and a score of others, all playing out the idea of the fleeting nature of life, and using the skull to remind us of the ultimate melancholy: the skull is what we shall all, one day, become. And it has been thus, ever since—right to the present day.

The Skull as a Warning

HOW DO YOU CREATE a sign proclaiming Warning! Danger! that will endure and be heeded for 10,000 years by whoever is around then to see it?

The United States Department of Energy has been pondering this problem for at least the last decade. The question first arose when it was decided to store, safely and secure from either accidental or deliberate human intrusion, enormous quantities of America's steadily accumulating stockpiles of high-level, highly dangerous nuclear waste, in an enormous depository to be dug deep below a remote range of mountains in southern Nevada.

The material, largely made up of tens of thousands of drums of uranium and plutonium isotopes, all by-products from the country's atomic power and nuclear weapons programs, would still have the capacity to kill and poison in hundreds of thousands of years. Hence the dilemma: how to persuade the people of the distant future—people in the unimaginable mists of Deep Time—that what is buried below the Nevada desert is now and virtually forever utterly and incomparably lethal, and should not under any circumstances be prodded, poked, or drilled down into. Ever.

A contest was held, and scores of strange ideas were advanced: forests of randomly sited concrete spikes; enormous black cement blocks; berms of solid salt; even local yucca plants genetically altered to a vivid blue color, to demonstrate the sinister powers of the buried materials (though critics said tourists in the unimaginable future, rather than being deterred, might flock to investigate a sky blue desert).

Three of the proposals—all moot for now, since President Barack Obama halted funding for the Nevada site in 2011, though it's generally agreed that such warning signs will be needed sometime soon, and somewhere— enjoyed a particular resonance.

One, the most macabre, suggested that people be allowed—encouraged, even—to wander near the site, so they would suffer the effects of powerful radiation and die, leaving their decaying bodies and scattered skeletons to warn others to stay clear. A second suggested plastering the region with gigantic monolithic heads, Easter Island-size, all sculpted into simulacra of Edvard Munch's unforgettably haunting painting *The Scream*. The third proposed to litter the desert with millions of fingernail-size silicon chips imprinted with the skull-and-crossbones image—the chips mingling with the dirt and increasing in volume as one approached closer to the site itself.

Maybe one of these ideas will eventually be chosen, or maybe some as yet unknown conceptual artist will come up with another idea of such dizzying elegance and simplicity as to trump the rest. But for now, these last three appear the most promising candidates—and central to each of them, whether it is the actual human skeletons, the wide-mouthed stone faces, or the printed silicon chips—is the universal image of warning: the human skull.

The fact that a skull is so easily and instantly recognizable—our ability to see minute differences in living faces has long translated to a similar facility for skull recognition—plays a central role in the ubiquity of skull symbolism: whether we use a skull to indicate poison, or mortality, or, as in Nevada, danger, we see it—and we get it immediately. The skull has become shorthand for an entire set of unpleasantnesses—and generally (not always, as we'll see in other chapters) we instinctively shy away from anything that displays a skull, or a decayed head. The skull is a device that speaks for the unspeakable, and it does so with particular efficiency.

The Chinese were probably the first to perfect the art of skull-display-as-warning, initially at the behest of the infamously cruel female emperor (China's only one), Wu Zetian, who created her own dynasty during the seventh century AD. She is remembered for innumerable acts of demonstrative cruelty; most notorious among them was her order to mount some twenty-two skulls of defeated rebels high on the walls overlooking the main gate of Luoyang, the country's then eastern capital city. They were mounted specifically to disincline any challengers from daring to oppose her rule—and the harshness of her message proved starkly effective: Empress Wu ruled without question or challenge for the relatively lengthy period of fifteen years, and only when a tidal wave of corruption charges against her coincided with a grave illness did she step down.

Skulls have performed similarly symbolic roles in Western cities, too. In Stockholm, would-be miscreants were persuaded to think twice by a row of skulls from executed criminals mounted above the South Gate in the walls of Gamla Stan, the old town. Similarly, Goethe quite probably saw scores of human skulls impaled on iron poles outside the city walls of Frankfurt am Main, and in London, tar-covered, gull-pecked skeletons of pirates were left to rot in the tides of Wapping, their heads sending a simple harsh message: don't.

Pirates, themselves, long employed the all-too-well-known image of a skull and crossbones to warn of their presence and of their menacing intent. They were not the first: the skull and crossbones, so very obviously implying death—and probably violent death at that—was first used by chemists to mark containers of noxious substances, such as cyanides and mercuric compounds, with a warning of their lethality.

But when pirates adopted the symbol, it took on a second meaning. It indicated, as

This 1725 print engraving of Stede Bonnet in Charles Johnson's *A General History of the Pyrates* is one of the earliest uses of the classic pirate skull and crossbones.

the pirates undoubtedly intended, that their ships, like dangerous chemicals, were to be warily regarded, that their crews could and willingly would deal out death if molested; but it also indicated that these same ships had great reserves of ill-intentioned power, that they could roam the high seas at will and plunder and destroy as the mood took them—in short, that they could be dangerously lethal even when you did your best to avoid them.

The crossover—the events that allowed skull-as-warning to evolve into skull-as-power-symbol—seems to have occurred in the late eighteenth century. The pirates may have been the first to employ, in this new sense, the symbol that was purloined or borrowed from the druggists' trade; but in a time a host of others, ranging from celebrated army units to motorcycle gangs, followed suit. Be careful, their skull emblems now signaled: we are strong, we

are careless of life, and we should not be treated lightly. We can, and if provoked most certainly will, deal you a lethally powerful hand.

The Skull Denoting Power

PIRATES—THOSE WHO, as the law has it, take a ship on the high seas from the possession or control of those lawfully entitled to it—have created havoc around the world's seas for as long as mankind has been sailing them. Firmly entrenched in folklore are the eye patch, the parrot perched on the shoulder, the disfiguring scar, the wooden leg or hook for a hand, and the slew of cruelly appropriate punishments, such as walking the plank—all of these define our popular view of pirates. And yet all these piratical images have been subordinated to the one universal symbol that stands for this scourge of the oceans: the black flag, long known as the Jolly Roger, adorned front and center with two crossed white thigh bones surmounted by a skull.

For four hundred years, ever since the beginnings of Atlantic trade and the almost immediately consequent Atlantic piracy, the haunting vision of this flag with its sinister grinning skull has sent shudders down the spines of the captains of any peaceful merchant ships, their passengers and crew.

It is a flag that bespeaks an awful, terrifying power. And lest any think the Jolly Roger suggestive of just a clutch of capital fellows who get sozzled on rum and like nothing more than to belly up to Caribbean bars, the adjective *Jolly* coming from the eternal smile of the skull, the name *Roger* taken to be a generality for maleness—the reality is much more sordid: it was a common piratical punishment, for instance, to gouge open a living captive's stomach, drag out his entrails, and nail them to the ship's mast, and then force him to dance backwards along the deck, running his guts out like a clothesline. Thus does the romantic image of the skull-bearing pirates begin to fade.

To be attacked by a pirate ship was a terrifying experience. The scenario had a certain routine to it. Under the steady press of the westerlies, the transoceanic cargo vessel, laden with treasure or trade goods,

The haunting vision of this flag with its sinister grinning skull has sent shudders down the spines of the captains of any peaceful merchant ships, their passengers and crew.

would be lumbering heavily east through steady seas of warm aquamarine, minding its own business. Suddenly, a suite of sails would appear on the horizon. A small sloop, traveling fast, would sweep swiftly into sight.

At a distance, it might be flying the flag of a friendly nation, but within sight or hailing distance, its crew would unfurl the infamous skull-adorned pirate flag. The sloop would then come alongside, its gunners firing warning shots across the bows and into the sails, ripping them to shreds, and would then tack wildly so that its own sails would begin to flap madly from the mast. The victim, slowed by its loss of sail power, would be forced to lower its own ruined canvas and come to a dead stop in the sea. Grappling hooks would then be thrown and hawsers drawn tight. As soon as bulwark smashed against bulwark, scores of heavily armed, wild-eyed young men would swarm over the rails.

They would be brandishing cutlasses and sabers and light axes. They'd slash at anyone showing the slightest resistance or disapproval. Some of the pirates would round up the crew and begin interrogating them, beating them, stabbing them, all too often eviscerating or strangling them—in one famous case, nailing a sailor's feet to the deck, whipping him with rattan canes, and then slicing his limbs from his torso before throwing his carcass to the sharks. Others would rummage through the ship's holds and cabins, searching for anything of value or interest. There might be gold aboard; there'd certainly be guns and powder; maybe skilled crewmen who could be forced or persuaded to join the pirate ship. And then, perhaps mounting a final violent assault on the passengers by way of a Parthian shot, they would all swarm back onto their own ship, detach the ropes, and slip rapidly away, soon passing over the horizon, leaving whoever remained of the passengers and surviving crew to limp away for refuge and repairs.

Such was the awe-inspiring, fear-producing power of the men who displayed the skull and crossbones.

It was a symbol later adopted by others, bent on more legitimate purposes. The Seventeenth Regiment of Light Dragoons, raised in Scotland in 1759, used the

The famous "Death or Glory" cap badge featuring a skull and a pair of crossed thigh bones.

SS uniforms used a variety of insignia and this Totenkopf (death's head) was used from 1934-1945.

skull as its emblem, with the motto "Or Glory" beneath its crossed bones. Later amalgamations led to the formation, first in the nineteenth century, of the Seventeenth Lancers, who took part in the tragically famous Charge of the Light Brigade; then, in 1922, of the cumbersomely named Seventeenth/Twenty-first Lancers, who'd become famous as one of the hardest of all the modern British army's cavalry units, usually deployed in only the toughest of situations, from Italy to Suez to Northern Ireland. All the while these soldiers wore a cap badge—actually never called that, but always referred to as "the motto"—of the death's-head skull. Those who wore the motto were known as "The Death or Glory Boys," and it is worth noting that there is a signal difference between their version and those sported by the pirates: while the pirates' emblems might look jolly, no one could ever accuse the soldiers' skulls of grinning. The eyes of their skulls were ablaze with hatred and venom. These were skulls worn by soldiers who meant business.

Armies elsewhere followed suit. In Germany the skull symbol was known as the Totenkopf, the deathman's head, and it was used by the Prussian cavalry, the Freikorps, and, somewhat inevitably, by various Nazi units—Hitler's personal bodyguard employed the death's-head, as did the Third Panzer Division of the feared Waffen-SS. The monstrous fear that both of these units engendered, the Reich leadership assumed, came in large part from their display of larger

and ever larger versions of the Totenkopf, an image of utter menace.

And not just in Germany. From the Black Brigades of wartime Italy to the Swedish Hussars; from the Portuguese Military Police to the Third Infantry in South Korea; from Estonian partisans, White Russians, and Chilean guerrillas to the men of the 100th Squadron of the British Royal Air Force, to an Australian weapons platoon, and to the officers and U.S. Marines of the Reconnaissance Corps Battalions: the skull and crossed femurs has been an ideal symbol of power projection—a warning still, perhaps, but a warning that says much more than "Poison," or "Radiation." It cries out, much more starkly, "Don't Dare Mess with Us."

As, of course, it says also for the men who belong to the world's Hell's Angel motorcycle clubs, to whose slogan Respect Few: Fear None is added, invariably and all around the planet, either a human skull trailing a wake of flames or the more usual skull-and-crossbones. Either device is designed to complement the aggressive and not infrequently law-breaking activities of what are generally known as the "one percenters" (a name taken from an assertion that "99 percent of motorcyclists are law-abiding"). It does so by serving to warn, in just the same way as did the symbol when employed by Atlantic Ocean pirates, British soldiers, or Estonian partisans, that terrific power is vested in those who bear the skull and crossbones, power that may be unleashed at any time—and woe betide anyone who might ever think otherwise.

Skulls in Art

WITH ITS LARGE EYE SOCKETS and grinning mouth and its smooth and shiny cranium, the human skull looks in most cases younger and fitter than the head that once covered it—a somewhat macabre example of the neoteny, of the forced youthfulness, that humans have long desired and venerated. So the skull, though seen initially and superficially as a symbol of death and mortality, offers these days, by way of its uncannily innocent appearance, a means of reminding one of youth and rebirth—one of the many subtle reasons that the human skull has been of such enduring fascination to artists ever since art itself began on cave walls tens of thousands of years ago.

But such subtlety emerged only in the later years. In earlier representations— on tombs, in memorial sculptures, in engravings—the imagery is invariably sinister rather than subtle: the macabre dominates. The horrific, the damned, the beyond redemption: these are the characteristics of the skulls and skeletons of the Middle Ages, for example. If a skull grins down at us, it does so with diabolical menace; if its eyeless sockets stare at us, as Fra Angelico's skulls do, from beneath the crucified Christ, they do so with reproof and harsh judgment. There's no beauty in early skull art: only fear, and horror, and melancholia.

But then we get to the sixteenth century, and things begin to change. We start to see what painters such as Ambrosius Holbein and Albrecht Dürer see: the skull as an entity of hitherto unappreciated beauty, a confection of bones and shapes and surfaces, of something that survives beyond the grave, an icon that serves as a reminder both of death and of life. We will look more closely at Ambrosius Holbein's younger brother Hans and his famously playful work *The Ambassadors* later in this section; but Dürer's pencil drawings of skulls, from the 1520s, make the point that by now artists were looking beyond the mere symbolism of the object, considering instead its imperfect beauty, the complexity of its facial bone surface allowing interplays of light and shadow, seldom considered before, now to be made the centerpieces of a new branch of art, no longer wholly macabre.

For centuries following this small revolution, the skull , at least in Western art, became so commonplace as to assume the role of cliché—numerous images of learned men portrayed sitting with their hands lightly touching the cranium of an ancestor, as if to remind us what they themselves are advised to remember, that life is fleeting and that human embodiment is merely a temporary privilege.

Though in the art of other cultures the skull remained a frightening topic, European and American artists soon began to represent it in a much lighter vein, ridding it of all notions gray and sad, investing it instead with wit and satire, albeit with an edge of the bizarre and the haunting.

Few will forget, for instance, Vincent van Gogh's fevered oil painting of a skull with a cigarette clutched between its teeth, nor can we overlook the images of skulls constructed of bricks and pipes, of naked female bodies (like those of Salvador Dalí), or rendered out of the shadows and reflections of a woman performing her toilette (as with the exquisite sketch by C. Allan Gilbert).

We laugh mildly at the identical monochrome skulls, presented by the score, in reproductions offered by Andy Warhol. We are (to judge from her sales figures) wildly enthusiastic about the works of Georgia O'Keefe, who saw beauty and poetry in the desert-dry animal skulls of New Mexico and

> ## Artists were looking beyond the mere symbolism of the object, considering instead its imperfect beauty, the complexity of its facial bone surface allowing interplays of light and shadow.

A Skull Sectioned was drawn by Leonardo da Vinci in 1489. It is one of his earliest anatomical drawings and is a detailed study of the skull, its dentition, proportions, and internal forms. The original is held in the Royal Library at Windsor Castle.

The Coat of Arms with the Skull, Albrecht Dürer, 1507. Engraving.

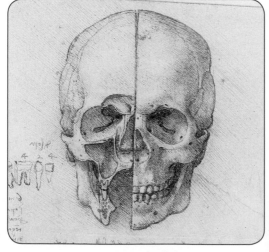

painted them relentlessly, framed by flowers that are, in life, as delicate and as fragile as the long-dead rostrum bones beside them. We are awed and repulsed by the skulls covered in diamonds, malachite, pearls—or dead flies—made by artist-jokesters like Steven Gregory and Damien Hirst (echoing, of course, the malachite–and jewel-encrusted skulls crafted by the Aztecs). And most of us are entirely repulsed by the work of a British artist who likes us to believe that she enjoys a special artistic enlightenment when she trepans her own skull with a Black & Decker drill.

Finally, not a few of us, are mildly amused by Grayson Perry, a British contemporary artist, who, in adopting skull art as his own, indicates that the skull has now become entirely demystified and demythologized by today's art and that if there ever was a time when a skull was a terrifying thing, before which one trembled, that time, so far as art is concerned, is long ago and far away.

The skull that was once all death and no subtlety is now all subtlety and certainly no death. Not anymore. For now, anyway.

Skull of a Skeleton with a Burning Cigarette, Vincent van Gogh, 1885 or early 1886. Oil on canvas.

Two Skulls in a Window Niche, by Hans or Ambrosius Holbein, 1520. Oil on limewood.

Holbein—The Ambassadors

AT FIRST BLUSH the most famous skull in art doesn't look like a skull at all. It looks more like a smear, a mistake, an inexplicable and rather unseemly interloper in the otherwise impeccably rendered and fully recognizable portrait of two great sixteenth-century men: *The Ambassadors*.

The rich complexity of Hans Holbein's enormous painting—it's almost seven feet square, and quite dominates the room in London's National Gallery where it has been housed since its acquisition in 1890—is little short of staggering.

First of all, there is the rich and powerful brace of ambassadors themselves. On the left of the picture is the ermine-clad, Falstaff-bearded, and precociously mature-looking (yet only 29 years old) French Ambassador to the court of King Henry VIII; he is the landowner Jean de Dinteville, the man who commissioned Holbein to create the picture. On the painting's right is his friend, the appropriately sobersided and clerically attired bishop of Lavaur, Georges de Selve, who was four years younger than de Dinteville but had already notched up an impressive diplomatic career as the French king's legate to the Republic of Venice, to Austria, to the Vatican, to Germany, Spain, and England. (He had been appointed to his bishopric, in the Languedoc, when he was just eighteen.)

Between the pair of men is a bewildering array of scientific equipment, all of which, later critics have surmised, holds symbolism of one kind or another. There is a globe of the then-known world, showing Rome standing proudly at its center. There's a plumb line: ancient Greek artists competed for who could draw the thinnest line, and here Holbein seems to be entering this contest once again, two millennia later, with a fine line neatly dividing the two ambassadors almost exactly halfway. Near the bottom of the oil-on-oak masterpiece is a lute with a broken string—depicted with another exceedingly fine line—that is taken to suggest discord (like the string's, a repairable discord) within the Church since the lute is next to a hymnbook that close study shows to be opened to a translation by the anti-Papist Martin Luther.

All this symbolism becomes much clearer when we spy the date of the picture. A precise time and date—perhaps of when Holbein started his work—is visible on a cylindrical sundial that stands almost in the picture's center: it can be read as depicting the moment at a little after 4 p.m. on April 11, 1533, Good Friday. And since this day was just a few weeks after the already-married Henry had wed Anne Boleyn, incurring the wrath of Pope Clement VII, the king's excommunication, and the eventual split between England and Rome, one might surmise that Holbein's entire painting, crammed with totemic imagery, is built around the turmoil that enveloped England, Europe, and the established Church of the day.

But where does the mysterious skull fit into all of this? That otherwise unrecognizable smear draped diagonally across the lower tier of the picture, tilted up from left to right, is very definitely a skull—or at least, it becomes one when the picture is viewed at a slant, either from its bottom left, or from the top right.

The distortion is an artistic conceit, a gimmick, a perspective-based jeu d'esprit, known as an *anamorphosis*, a form that has enjoyed occasional popularity in the painterly world. Leonardo da Vinci and Albrecht Dürer, for instance, employed the technique in some of their more frivolously experimental works. Today one seldom sees it in serious art, any more than one sees trompe l'oeil. Yet often we spy anamorphoses on our television screens during sporting events, since they're often used to project advertisements onto a football field or a rugby pitch. In full view they appears as gibberish, as does the Holbein skull when seen from straight on; but when viewed from a certain angle, by the fans in a particular stand, say, all is revealed as Vodafone, or Verizon, or whoever has paid for this costly kind of promotion.

Elaborate mathematical formulas have been contrived to establish with precision where best—from which side, and at which height—to view the Holbein image (many insist that you would see the skull when ascending a staircase at the top of which the enormous picture was intended to be hung). When the viewer positions himself correctly, whether casually on the stairs or otherwise by dint of geometry to one side, the skull magically appears, the left side of its cranium brightly lit, the right eye socket somewhat larger than the left, the teeth mostly missing. It's not a handsome skull—it would not find favor with Damien Hirst, nor would it turn up in a good museum. It's more grotesque than elegant.

And yet in being so, and in being disguised from the casual viewer, it serves the purpose for which the artist included it: to remind the careful student of the absolute certainty of mortality. Indeed, reminders of the inevitability of life's end are common in the art of the sixteenth century (and it's worth pointing out that extremely careful scrutiny of this painting shows another tiny skull, distorted beyond all recognition, lurking in de Dinteville's fur hat, presumably as a further reminder).

But who decided to include it? What trumps most considerations is de Dinteville's family motto: Memento mori—Remember we die. This adds to the probability that the conceit was not Holbein's, that he was instructed to include it by a somewhat surprisingly humble, realistic, and rather wise ambassador, who knew it to be appropriate and proper to make a public declaration of his awareness that his worldly existence would one day come to an end.

And what better way to do so than by placing in an otherwise supremely realistic—and yet symbol-filled—painting, an elaborately encoded inclusion of the most recognizable image of mortality: a human skull? It was a stroke of artistic genius that has intrigued and delighted audiences for nearly 500 years—a feat that even Damien Hirst and his *For the Love of God* may have some difficulty in matching.

The Ambassadors, Hans Holbein, 1533. Oil on oak.

The Skull in Mexico

NOWHERE ELSE IN THE WORLD has the human skull been so comprehensively integrated into local culture as in Mexico. The Mexican people have an almost jovial familiarity with death and its various insignia, especially the *calavera*—the skull. Absent are the various denials of mortality so common in European cultures: instead, just a comfortable intimacy, peculiarly Mexican, with the inevitable. Mexicans, not more generally Latin Americans: no other people south of the Rio Grande appear to have, or ever to have had, the kind of obsessive interest in man's passage out of life that is so characteristic of the Mexicans.

Nor is this in any sense recent. It's all too easy to assume that the wild celebrations of death staged countrywide at the end of October and the beginning of November each year—notably those on the Día de los Muertos (Day of the Dead)—are just a strange Mexican take on Christian enthusiasms imported by the Spaniards six centuries ago.

But although in modern times Mexicans have imported European death symbols (Black Death dances, tarot card images, that kind of thing) into their traditional ceremonies, it would be wrong to assume, despite their Spanish name, an Hispanic origin for Mexican death feasts. The skull and other macabre icons of mortality had been appearing in Mexican art for perhaps a millennium before the Europeans arrived. Rattling skeleton bones are firmly annealed into every one of the tribal components of Mexican culture.

Human heads are to be found in the art of the Maya of Yucatán and Chiapas. Cranial iconography is a central part of the canon of the ancient Zapotecs and Mixtecs of Oaxaca. The grave and its occupants are crucial to the rituals of the Toltecs and the Michoacán of the country's mid-regions. Skulls—not only of human bone, but carved also from turquoise and obsidian (though probably not, as we shall see, from Indiana Jones–style rock crystal), and even of spun sugar and pastry, presented as toys and playthings made to be eaten—have been part and parcel of the life of those Aztecs who now call themselves the Mexica and whose capital city, Tenochtitlán, is now called Mexico City (the urban giant that is still the center of the curious world of Mexican skull worship).

There has never been a satisfactory explanation: Why is Mexico so different? In both Mayan and Aztec mythologies there are innumerable death gods—the Mayan tradition being especially complicated: gods that hunt, gods that bring specific lethal diseases, gods disguised as jaguars. More than a million Mexicans speak Mayan dialects and continue to live where death arises regularly from the dangers of the jungle (snakebites, animal attacks, cave-ins). Perhaps, after all, it is not surprising that the traditions of centuries have survived, death riven firmly into Mayan culture.

The Aztec association with death is rather less complex, perhaps more readily understandable as a component of today's rituals—not least because the great central cities of modern Mexico display so many Aztec remains. The chief Aztec death god is Mictlantecuhtli—the Lord of the Underworld, depicted in thousands of

In Mexican folk culture La Catarina is the skeleton of a high society woman and one of the most popular figures of the Day of the Dead celebrations.

ancient statues and icons and living on in depictions and drawings to this day (including a popular comic book series, *Mictlantecuhtli, the Aztec Zombie*). The god is utterly macabre: he is usually portrayed as having a blood-spattered skull, a headdress of owl feathers and paper strips, a necklace of eyeballs, bone discs in his earlobes, and skeleton feet incongruously protected by sandals. He sits, he grins toothlessly, he presides, laughing all the while over the bizarreries of the world that Quetzalcoatl created and over whose millions of souls he and his brother deities are presumed to rule, deep in their alternative underworld.

It's thus perhaps none too surprising that two of the world's best-known ancient skulls have come from death-obsessed ancient Mexico—or at least, were thought to have done so. Both are in the British Museum. Both are extraordinary and beautiful. But only one of them, it seems, is a genuine Aztec creation.

The first, the real one, is amazingly intricate: it's made of a real human cranium, lined with deerskin. Deerskin ligaments secure the mandible to the cranium in such a way as to allow the jaws to open and close. The skull is covered with mosaics arranged in bands—a band of bright blue turquoise tiles, followed by a band of black lignite tiles, then more turquoise, then yet more lignite—five bands in all. The eyes are polished spheres of iron pyrite, set into rings of polished white conch shell. And if this weren't sufficient splendor for the priests who would wear it (attached to their dress with more straps of deerskin, according to drawings found in Aztec documents), its nose socket is lined with plates of shell from a sea creature known as the spiny oyster. That shell, incidentally, like the pyrite and the conch and the lignite, cannot be found where the Aztec craftsmen made this exquisitely memorable object. These materials came from hundreds of miles away.

The other skull is a thing of stunning beauty—and, thanks to Hollywood and

The mosaic skull from the British Museum is believed to represent the god Tezcatlipoca, or "Smoking Mirror," one of four powerful creator deities who were among the most important gods in the Aztec religion. It dates from the fifteenth to sixteenth century and is on display in the Central and South American section of the museum.

the Indiana Jones franchise, an object of bemusement and veneration to millions. It is a life-sized transparent skull carved from a single block of rock crystal. Held in the hand and viewed from shifting angles, it is a marvel to behold: its internal flaws and reflections, its smooth glass-like cranial surface, the sturdy precision of its jawline, the cunning optical illusions caused by its incarvings (the eye sockets) and notchings (the teeth)—the total effect is just overwhelming.

The British Museum bought the crystal skull from Tiffany in New York in 1897, assuming that it was indeed an Aztec representation, as ancient as the basalt and limestone skulls that adorn the great Aztec temples in Mexico City, carved just as they had been, probably just as central to the native priestly ritual.

But no. Modern examination of the object has shown that the crystal is from

Brazil, far too far away from the Aztec trade routes, and that the instruments employed to carve it were almost certainly grinding wheels and hard abrasives—not the hand tools used to carve genuine Aztec sculptures. The skull is a clever fake, certainly not made in Aztec times, and maybe not made in Mexico at all.

Much the same conclusion was reached when traces of a synthetic grinding agent, carborundum, were found in crevices between the teeth of other crystal skulls—notably a translucent skull with a hollow cranium, purportedly bought in Mexico City in 1960, and sent to the Smithsonian Institution in Washington in 1992. The inescapable conclusion is that it, too, is of relatively modern manufacture.

Common to these forgeries—the two aforementioned and three others, small crystal skulls now on display at a museum in Paris—is a lone French antiquities dealer

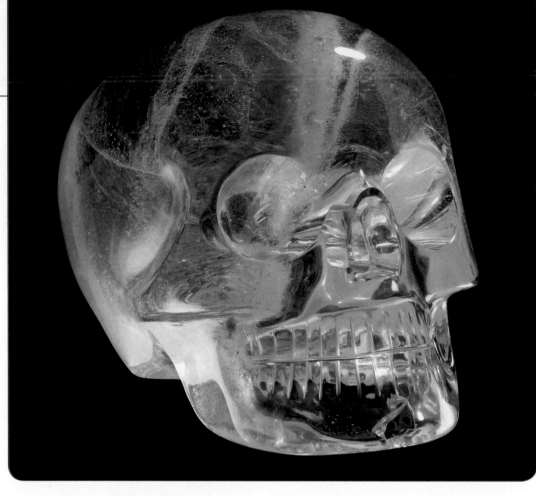

named Eugène Boban, who was chief archaeologist in the Mexican court of the Emperor Maximilian. A flamboyant figure, he had a stall at the Paris Exposition of 1867 and later ran a dealership in Manhattan, selling objets trouvés from his various expeditions into the Mexican jungles—often selling with more enthusiasm, it now seems, than integrity.

As to why this Aztec and Mayan fascination with death: One can only suppose that the terrors of the earthquakes and volcanoes to which Mexico is so prone, together with its history of economic harshness and political savagery, have much to do with it. Certainly it is true today, with drug wars raging in the country's north, with popular revolutions raging and petering out betimes, with ground shaking and lava pouring and glowing ash clouds occurring regularly, that one must see in Mexico a certain need for fatalism. But why more so here than in similarly afflicted places—Kashmir, say, or Sudan or the Congo—there seems no obvious reason.

Whatever the anthropological explanation, the legacy of death worship remains, and most notably during the delightful madness of the Days of the Dead, holidays staged each year beginning around the end of October and culminating in the Catholic remembrance days of All Saints (November 1) and All Souls (November 2). In almost every conceivable manner—parties, dances, feasts, rituals—Mexicans use the period to memorialize the souls of those who have died, reserving particular days for those who died violently, those who died as children, those who passed away in old age, those who died unbaptized, those who are forgotten.

The great Mexican artist-printmaker José Guadalupe Posada (who died in 1913) was for decades the artist laureate of the celebration, images of *calavera* featuring in almost all of his fantastical cartoons. His dancing, smoking, singing, laughing, guitar-playing, horse-riding, gun-totin', tequila-drinking, preaching , praying, fighting, and love-making skull figures are everywhere still, part of the backdrop to a celebration that is unique in all the world.

Skulls. Skulls by the million: skulls that are usually made of spun sugar, and used as offering to the departed—a remembrance to the sweetness of life and the reality of death. Invariably brightly colored, adorned with tinsel and marigold petals, their smiles often tricked out in contrasting colors, broad and wide, more than any other object seen in today's Mexico, the confections that proliferate during these feasts are the skull as an object of good cheer and inevitability: there is none of the grim association that the skull has elsewhere.

In Mexico, on All Souls' Day, you dance, you down a glass or two of tequila or Dos Equis beer, and you bite hard into an intensely sweet sugar skull. Thereby, you revel in one of the inescapable facts of life, contemplating eternal happiness in your future and giving thanks for times past. As aide mémoire, there can surely be no better device, say all Mexicans, than a human skull like this, made prettily by hand, out of rainbow-colored candy. All hail, they say, to the Mayans and the Aztecs, and the long ago Gods of Death!

Mammals

Aardvarks

Armadillos

Aardvark
Orycteropus afer ◁

THE AARDVARK'S SKULL is characterized by a long, narrow snout (or rostrum, to give it its correct term) that allows it to reach into ant and termite nests. The rostrum houses a long, sticky tongue that can probe into crevices. In the wild, the burrows an aardvark creates with its long front claws and strong limbs are used by a whole host of other animals. It therefore plays an important ecological role in many habitats.

AKA: Ant Bear
Kingdom: Animalia
Phylum: Chordata
Class: Mammalia

Order: Tubulidentata
Family: Orycteropodidae
Genus: Orycteropus
Behavior: Insectivore/Nocturnal

Six-banded Armadillo
Euphractus sexcinctus ▷

THE SKULL OF THE six-banded armadillo is displayed here both with and without the armored plate that sits atop its head. The armor is made up of scalelike scutes of horny skin. In its native South America, it feeds on grubs, insects, and plants. It can roll into a ball for protection when threatened.

AKA: Yellow Armadillo
Kingdom: Animalia
Phylum: Chordata
Class: Mammalia
Order: Xenarthra

Family: Dasypodidae
Genus: Euphractus
Behavior: Omnivore/
Diurnal

Nine-banded Armadillo
Dasypus novemcinctus ◁

THE NINE-BANDED armadillo is a common road kill in North America—a boon to any skull collector in that part of the world. This specimen retains its distinctive head carapace, which is formed from ossified dermal scutes. This species cannot roll into a ball, as the three-banded armadillo can.

AKA: Nine-banded
Long-nosed Armadillo
Kingdom: Animalia
Phylum: Chordata
Class: Mammalia

Order: Xenarthra
Family: Dasypodidae
Genus: Dasypus
Behavior: Insectivore/
Nocturnal

Long-nosed Bandicoot
Perameles nasuta △

THE LONG-NOSED BANDICOOT is a marsupial, with, as its name suggests, a long rostrum, which has an interesting flare at its end. The bandicoot is insectivorous and relies on its sensitive long nose and well-developed sense of smell to seek out prey.

Kingdom: Animalia
Phylum: Chordata
Class: Mammalia
Order: Peramelemorphia

Family: Peramelidae
Genus: Perameles
Behavior: Omnivore/
Nocturnal

Southern Brown Bandicoot
Isoodon obesulus ▷

IT SEEMS A BIT of a misnomer to describe this skull as coming from a short-nosed animal, but that's what it is. The southern brown bandicoot belongs to the "short-nosed bandicoot" genus—and Dudley even has a long-nosed bandicoot to compare it to. This bandicoot species, like many others, is threatened in its native Australia by introduced predators such as the red fox. However, in some areas populations appear unaffected; bandicoots can sometimes be seen in suburban Adelaide.

AKA: Quenda
Kingdom: Animalia
Phylum: Chordata
Class: Mammalia
Order: Peramelemorphia
Family: Peramelidae
Genus: Isoodon
Behavior: Omnivore/Nocturnal

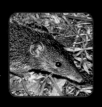

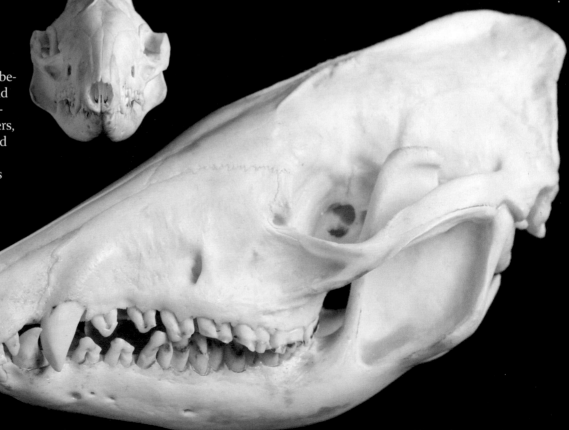

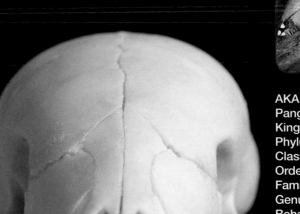

Tree Pangolin
Manis tricuspis

THIS STRANGE, FEATURELESS skull seems to be missing its lower jaw. Pangolins like this tree-dwelling species are known to roll into a protective ball when threatened. The tree pangolin is generally nocturnal, and a specialist insect eater. The apparent lack of any teeth is explained by the fact that it uses its sixty-centimeter (twenty-inch) adhesive tongue to gather its food. The body is covered in hard scales to protect it from attacks from ants when it uses its sharp claws to dig out their nests.

AKA: White-bellied Pangolin
Kingdom: Animalia
Phylum: Chordata
Class: Mammalia
Order: Pholidota
Family: Manidae
Genus: Manis
Behavior: Insectivore/ Nocturnal

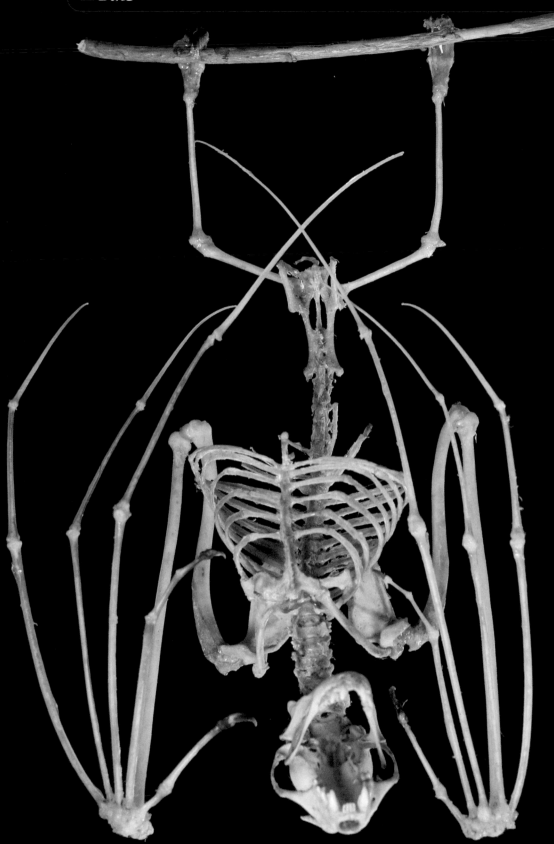

Lesser Short-nosed Fruit Bat
Cynopterus brachyotis

MOST SMALLER BATS use echolocation to navigate, but the large orbits of this Southeast Asian bat tell us that it uses its eyes to navigate. It's a frugivore as its name suggests but, nonetheless, has an impressive set of teeth. The skull here is mounted on a full skeleton.

Kingdom: Animalia
Phylum: Chordata
Class: Mammalia
Order: Chiroptera
Family: Pteropodidae
Genus: Cynopterus
Behavior: Frugivore/ Nocturnal

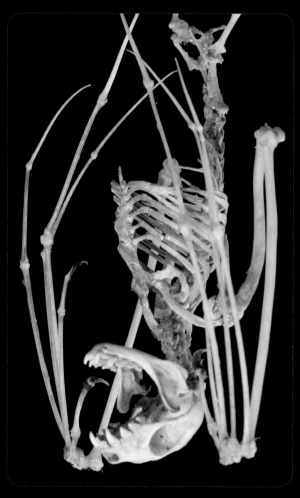

Hammer-headed Bat
Hypsignathus monstrosus

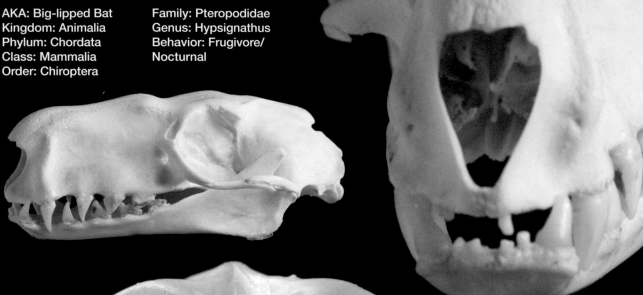

THIS IS A LARGE fruit-eating bat with a 0.91-meter (three-foot) wingspan, a large head, and a unique skull with sharklike teeth. The enlarged rostrum allows this bat to produce loud, far-carrying honks.

AKA: Big-lipped Bat
Kingdom: Animalia
Phylum: Chordata
Class: Mammalia
Order: Chiroptera

Family: Pteropodidae
Genus: Hypsignathus
Behavior: Frugivore/
Nocturnal

Javan Tailless Fruit Bat
Megaerops kusnotoi

THE JAVAN TAILLESS FRUIT BAT, like the other fruit eaters, has large, flat-topped teeth, quite un-like the small pointed teeth of its insectivorous cousins. The skull is large and resembles—as its more common name "flying fox" suggests—a fox or even a small dog. Its skull is fine to the point of translucence, a feature of all bat skulls.

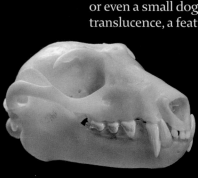

Kingdom: Animalia
Phylum: Chordata
Class: Mammalia
Order: Chiroptera
Family: Pteropodidae
Genus: Megaerops
Behavior: Frugivore/Nocturnal

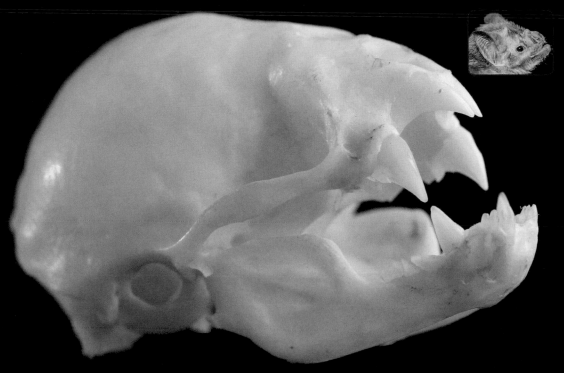

Common Vampire Bat
Desmodus rotundus

THIS SKULL IS TINY, delicate, and paper thin with disproportionately large teeth. The common vampire bat has teeth so sharp that you cannot feel its bite, and anticoagulants in its saliva keep you bleeding until it has had its fill. There are three species of vampire bat, each sufficiently different from the others to warrant a separate genus.

Kingdom: Animalia
Phylum: Chordata
Class: Mammalia
Order: Chiroptera

Family: Phyllostomidae
Genus: Desmodus
Behavior: Carnivore/
Nocturnal

DUDLEY'S NOTES:
Fang-tastic! This vampire bat is unlike any other bat I've ever seen in that the two front teeth actually come together as a cutting tooth to pierce the skin.

Bicolored Leaf-nosed Bat
Hipposideros bicolor

THE BICOLORED LEAF-NOSED BAT is widespread in Australia and Southeast Asia. The enlarged nasal passages are surrounded by flaps of skin that the bat uses to assist with echolocation. Bicolored leaf-nosed bats can sometimes be bleached a bright orange by the pungent fumes found in their caves.

AKA: Bicolored Roundleaf Bat
Kingdom: Animalia
Phylum: Chordata
Class: Mammalia
Order: Chiroptera
Family: Rhinolophidae
Genus: Hipposideros
Behavior: Insectivore/
Nocturnal

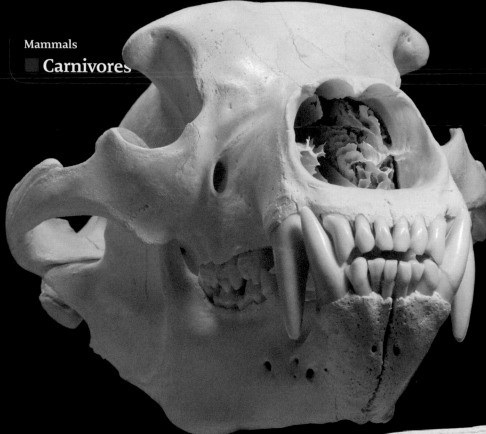

Grizzly Bear
Ursus arctos horribilis

THE GRIZZLY BEAR has a large, powerful and impressive skull to grace any collection. Grizzly bears are omnivorous and will adapt their diet seasonally to take advantage of abundant food sources. The bears of Yellowstone National Park feast annually on calorie-rich moths.

AKA: North American Brown Bear
Kingdom: Animalia
Phylum: Chordata
Class: Mammalia

Order: Carnivora
Family: Ursidae
Genus: Ursus
Behavior: Omnivore/ Nocturnal

Polar Bear
Ursus maritimus

THE POLAR BEAR is an impressive carnivore, and of all the bears, its skull is the most sought after. Like all bear skulls, it is elongated with a distinct gap between the carnassials and the canines.

Kingdom: Animalia
Phylum: Chordata
Class: Mammalia
Order: Carnivora
Family: Ursidae
Genus: Ursus
Behavior: Carnivore/ Diurnal

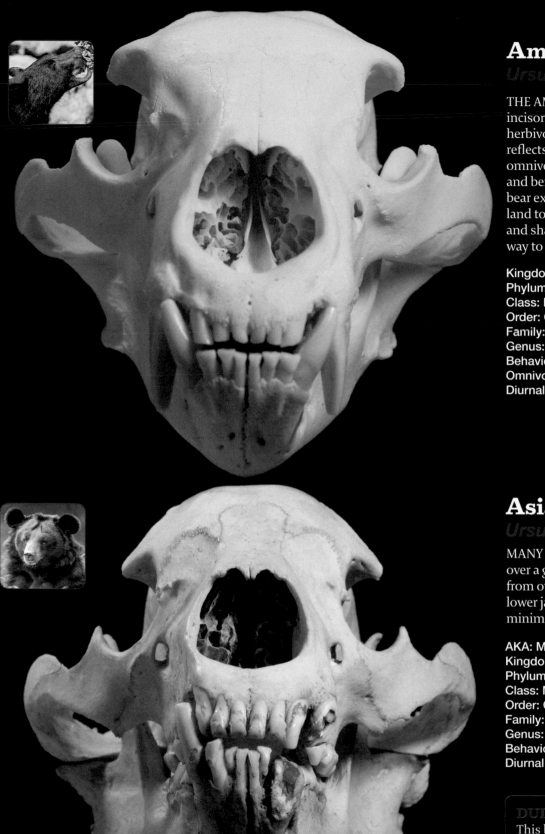

American Black Bear
Ursus americanus

THE AMERICAN BLACK BEAR has the canines and incisors of a typical carnivore but the molars of a herbivore. This specialization of different teeth reflects the fact that American black bears are omnivorous and feed mainly on a variety of plants and berries. Many subspecies of American black bear exist (they occur from Mexico to Newfoundland to Alaska). There is a wide variety in skull size and shape. However, their dental formula is a sure way to identify them.

Kingdom: Animalia
Phylum: Chordata
Class: Mammalia
Order: Carnivora
Family: Ursidae
Genus: Ursus
Behavior:
Omnivore/
Diurnal

Asiatic Black Bear
Ursus thibetanus

MANY SUBSPECIES of the Asiatic black bear exist over a great range. Its skull can be distinguished from other bears' skulls by the particularly massive lower jaw, much narrower zygomatic arches, and a minimal sagittal crest.

AKA: Moon Bear
Kingdom: Animalia
Phylum: Chordata
Class: Mammalia
Order: Carnivora
Family: Ursidae
Genus: Ursus
Behavior: Omnivore/
Diurnal

> **DUDLEY'S NOTES:**
> This bear is unusual because it was a captive-bred animal and its teeth show evidence of dentistry work. The front canine has what appears to be a lead filling, which I've never seen before in any skull I've ever owned.

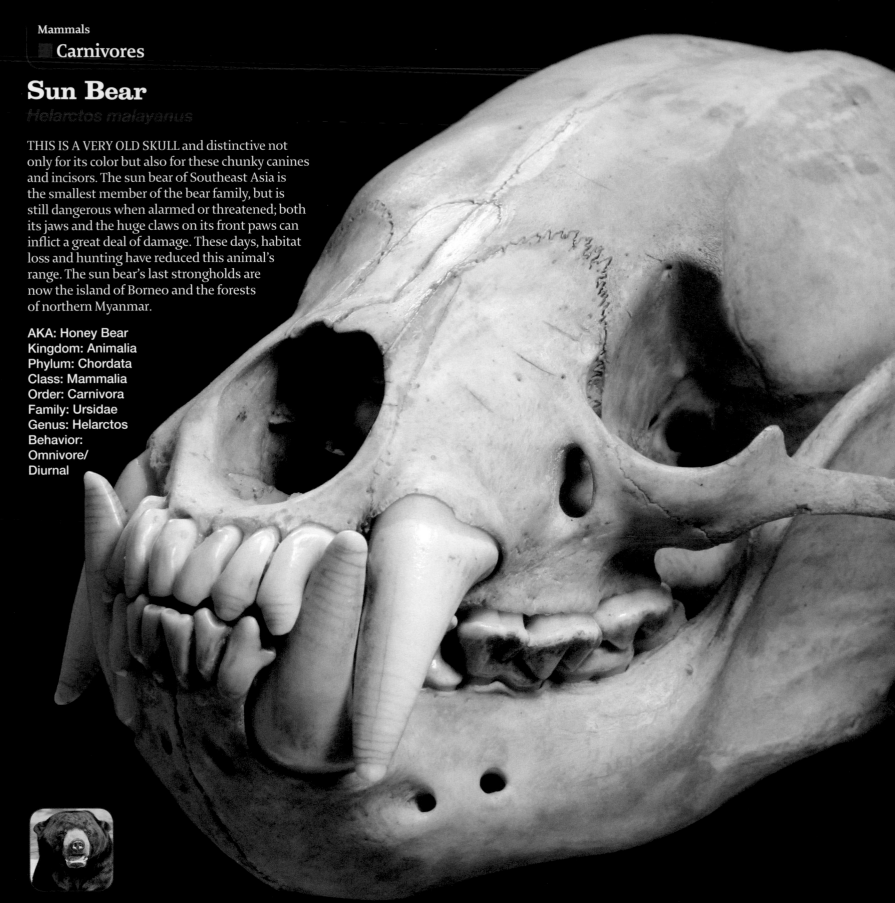

Sun Bear
Helarctos malayanus

THIS IS A VERY OLD SKULL and distinctive not only for its color but also for these chunky canines and incisors. The sun bear of Southeast Asia is the smallest member of the bear family, but is still dangerous when alarmed or threatened; both its jaws and the huge claws on its front paws can inflict a great deal of damage. These days, habitat loss and hunting have reduced this animal's range. The sun bear's last strongholds are now the island of Borneo and the forests of northern Myanmar.

AKA: Honey Bear
Kingdom: Animalia
Phylum: Chordata
Class: Mammalia
Order: Carnivora
Family: Ursidae
Genus: Helarctos
Behavior:
Omnivore/
Diurnal

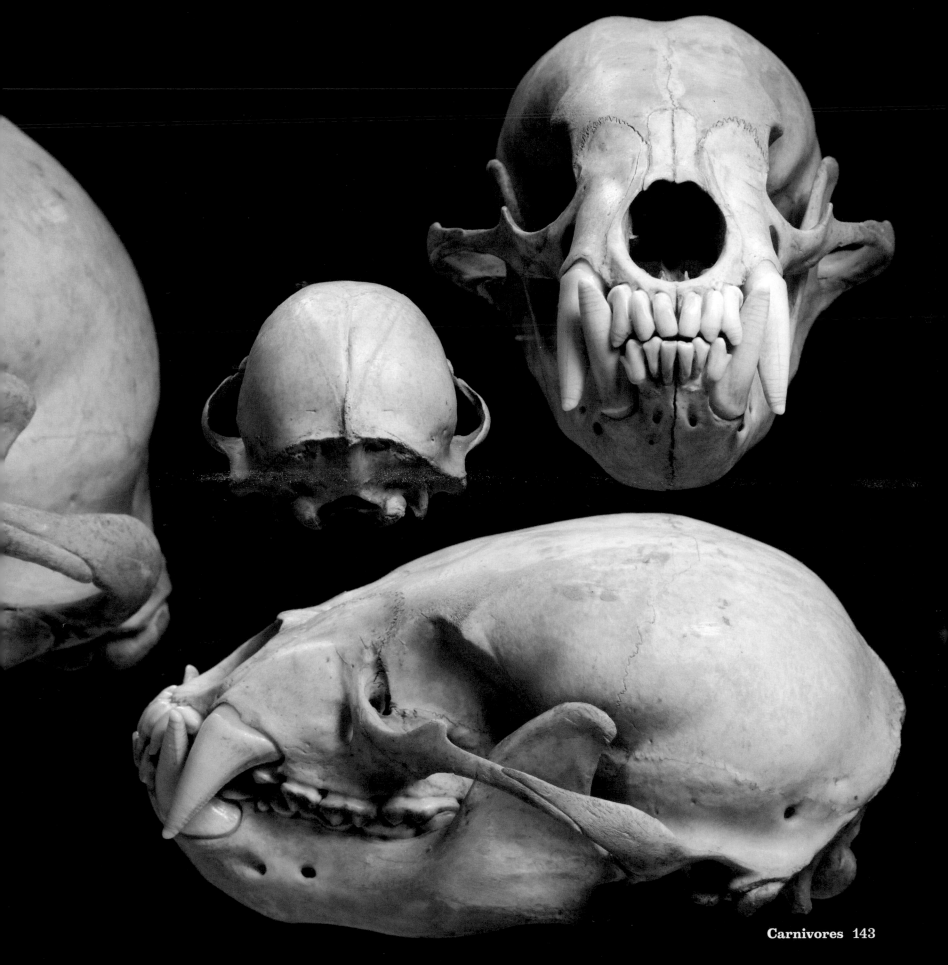

African Lion

Panthera leo

THE AFRICAN LION has a wonderfully powerful skull. The "king of the beasts" is rivaled for its crown only by the tiger. The skulls are all but indistinguishable, such is the similarity between the two species. Some argue that tigers have comparably wider muzzles but that lions have wider zygomatic arches and larger canines. Lions and tigers can interbreed to produce supersized but thankfully sterile offspring known as ligers or tigons (depending on parentage).

Kingdom: Animalia
Phylum: Chordata
Class: Mammalia
Order: Carnivora
Family: Felidae
Genus: Panthera
Behavior: Carnivore/Diurnal

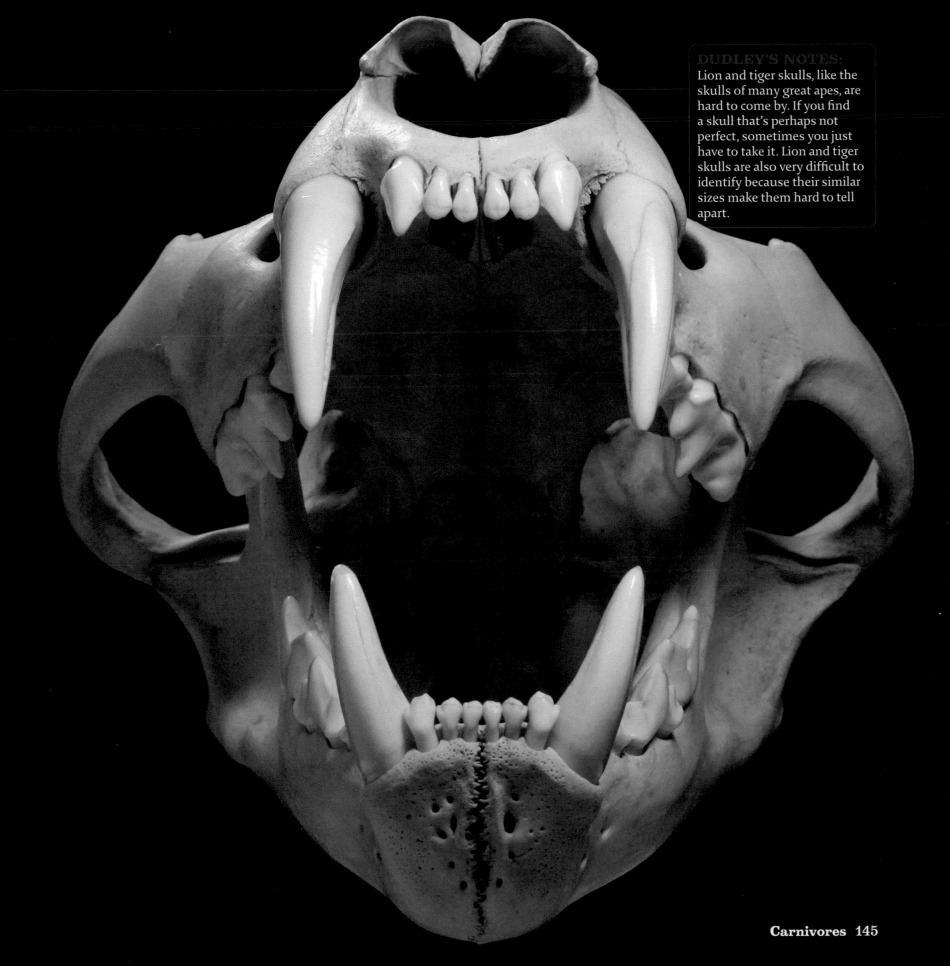

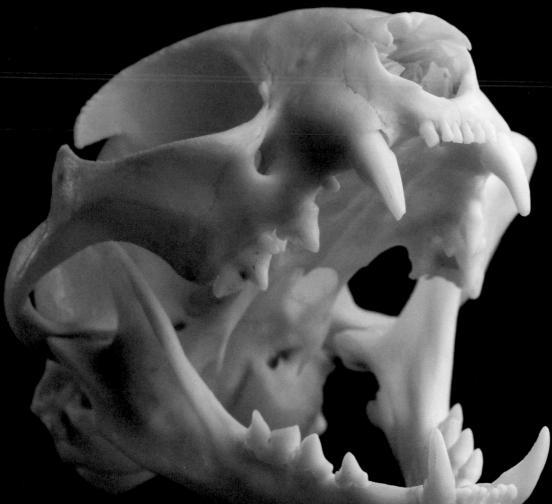

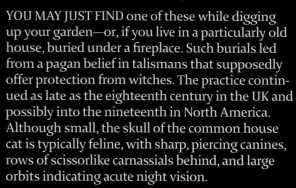

Domestic Cat
Felis catus

YOU MAY JUST FIND one of these while digging up your garden—or, if you live in a particularly old house, buried under a fireplace. Such burials led from a pagan belief in talismans that supposedly offer protection from witches. The practice continued as late as the eighteenth century in the UK and possibly into the nineteenth in North America. Although small, the skull of the common house cat is typically feline, with sharp, piercing canines, rows of scissorlike carnassials behind, and large orbits indicating acute night vision.

Kingdom: Animalia
Phylum: Chordata
Class: Mammalia
Order: Carnivora

Family: Felidae
Genus: Felis
Behavior: Carnivore/
Diurnal

Persian Cat
Felis catus

THE PERSIAN CAT has been selectively bred to produce a large skull with a tiny rostrum, resulting in jaws that do not meet properly. This skull structure, known as brachycephaly, has resulted in a number of health problems in the breed.

Kingdom: Animalia
Phylum: Chordata
Class: Mammalia
Order: Carnivora

Family: Felidae
Genus: Felis
Behavior: Carnivore/
Diurnal

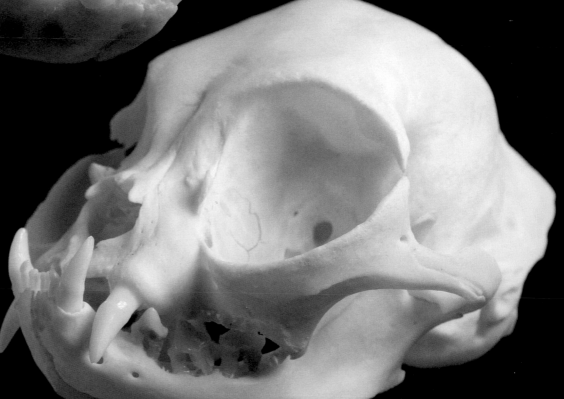

Clouded Leopard
Neofelis nebulosa

THE SMALL CLOUDED LEOPARD has the impressive distinction of possessing the longest canine teeth, relative to skull size, of any feline—even the tiger. It has a relatively small skull for a big cat. It is now split into two species—one in Sumatra and Borneo, the other across India, the Himalayas, and Indochina.

Kingdom: Animalia
Phylum: Chordata
Class: Mammalia
Order: Carnivora
Family: Felidae
Genus: Neofelis
Behavior: Carnivore/ Nocturnal

Leopard
Panthera pardus

BIG EYES, POWERFUL TEETH and jaws—this skull has killing machine written all over it. Leopards are opportunistic hunters and will take a wide variety of prey. They are particularly good climbers and will often drag their kill up a tree for later consumption.

Kingdom: Animalia
Phylum: Chordata
Class: Mammalia
Order: Carnivora
Family: Felidae
Genus: Panthera
Behavior: Carnivore/Nocturnal

Cheetah
Acinonyx jubatus

FOR A CARNIVORE, the cheetah has a very small head relative to its body size. The skull is very distinctive: its jaws are not as powerful as those of other big cats and have fewer and smaller teeth. Cheetahs cannot easily kill their prey by piercing wounds, as do leopards, but suffocate their prey by compressing the windpipe. They also lack the gap between the premolars and canines seen in other felines. It's all a sacrifice for speed that natural selection has produced in this spectacular and unique big cat.

Kingdom: Animalia
Phylum: Chordata
Class: Mammalia
Order: Carnivora
Family: Felidae
Genus: Acinonyx
Behavior: Carnivore/ Diurnal

Mountain Lion
Puma concolor

ALSO KNOWN AS A puma or cougar, the mountain lion has a skull that is short and wide, like those of all felines. It can be distinguished from the skulls of other big cats by the parietal bone, which exhibits long, fingerlike processes that reach forward over the frontal bone on each side of the skull. It has large orbits relative to skull size, an indication of how important vision is to this predator's style of hunting. The auditory bullae are large, suggesting that sound is also very important.

AKA: Cougar, Puma
Kingdom: Animalia
Phylum: Chordata
Class: Mammalia

Order: Carnivora
Family: Felidae
Genus: Puma
Behavior: Carnivore/Nocturnal

Saber-toothed Cat
Smilodon californicus ▷

REPLICA SKULLS, such as this saber-toothed cat, are available online. This one was based on a fossil from the tar pits of La Brea. The saber-toothed cat has a long and pronounced sagittal crest. In its current pose, the jaw, although wide, is not fully open. In fact, the jaw could be opened to 120 degrees to enable the cat to bite its prey. The zygomatic arches of this skull are much smaller than those of modern big cats such as the tiger (below).

AKA: Smilodon
Kingdom: Animalia
Phylum: Chordata
Class: Mammalia

Order: Carnivora
Family: Felidae
Behavior: Carnivore/
Nocturnal

Tiger
Panthera tigris

TIGERS ARE THE LARGEST of the extant big cats. The skull here is very similar to that of the lion elsewhere in the collection and has fearsome canines that can reach up to ten centimeters (four inches) long. There were nine subspecies of tiger; three (Javan, Bali, and Caspian) are now extinct, and the remaining ones are endangered. It is said that today there are more tigers in captivity in the United States than remain over the whole of their historic range in the wild. It is a rare and sought-after skull, as tiger bones are used in Chinese medicine.

Kingdom: Animalia
Phylum: Chordata
Class: Mammalia
Order: Carnivora
Family: Felidae
Genus: Panthera
Behavior: Carnivore/
Crepuscular

DUDLEY'S NOTES:
Tiger skulls are obviously very hard to get a hold of due to the fact that they're endangered. This one came from the taxidermist who gave me the lion.

Boston Terrier
▷ *Canis lupus familiaris*

THIS SKULL IS an instructive
example of the results that
selective breeding can have on a
dog skull. The rostrum is almost
completely absent, and the lower
jaw extends beyond the upper
and almost curls up above it.

Kingdom: Animalia
Phylum: Chordata
Class: Mammalia
Order: Carnivora

Family: Canidae
Genus: Canis
Behavior: Omnivore/
Diurnal

DUDLEY'S NOTES:
This species must have
tremendous difficulty breath-
ing because of its nose being
crushed, a direct result of
crossbreeding.

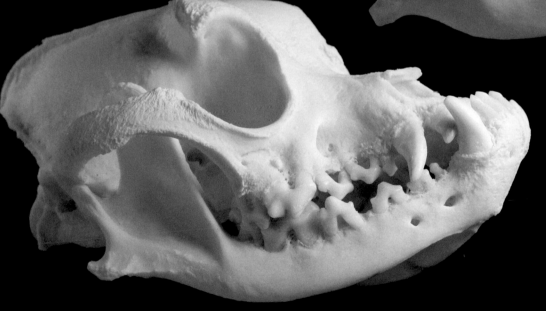

Boxer
◁ *Canis lupus familiaris*

LIKE THE BOSTON TERRIER, the boxer has been
bred for a short rostrum. The mandible extends
well beyond the upper jaw so that its bottom front
teeth fail to meet the top. Fortunately for the dog,
its carnassials do meet; otherwise it would have
problems eating anything other than soft dog food.

Kingdom: Animalia
Phylum: Chordata
Class: Mammalia
Order: Carnivora

Family: Canidae
Genus: Canis
Behavior: Omnivore/
Diurnal

Chihuahua
▷ *Canis lupus familiaris*

OF ALL THE DOG SKULLS in this collection, this one must surely be
the most extreme—both in terms of size and deformity. The Chihuahua is, effectively, a dwarf dog, retaining many dimensions we would
associate with a puppy, such as a very short, underdeveloped muzzle
and large cranium.

Kingdom: Animalia
Phylum: Chordata
Class: Mammalia
Order: Carnivora

Family: Canidae
Genus: Canis
Behavior: Carnivore/
Diurnal

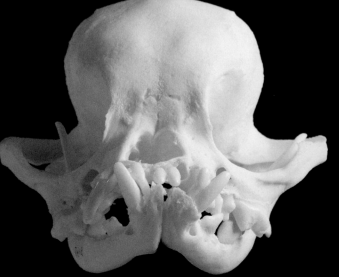

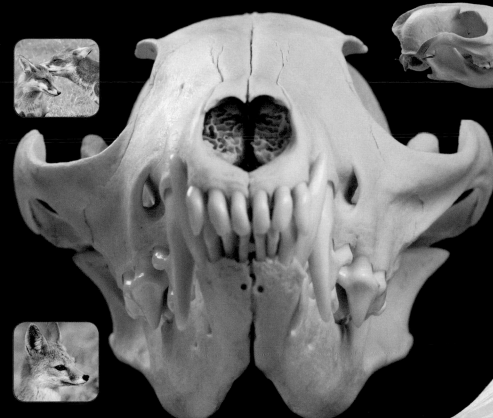

European Red Fox
◁ *Vulpes vulpes*

FOXES HAVE A TYPICALLY doglike skull exhibiting a classic combination of carnassial teeth, canines, incisors, and a sagittal crest. Foxes are widely distributed around the Northern Hemisphere, and have been introduced to Australia.

Kingdom: Animalia
Phylum: Chordata
Class: Mammalia
Order: Carnivora

Family: Canidae
Genus: Vulpes
Behavior: Carnivore/
Nocturnal

Kit Fox
▷ *Vulpes macrotis*

THIS DIMINUTIVE NOCTURNAL FOX of the south-western United States and northern Mexico has a delicate, small, but typically canine skull. The kit fox is only slightly bigger than a domestic cat.

Kingdom: Animalia
Phylum: Chordata
Class: Mammalia
Order: Carnivora

Family: Canidae
Genus: Vulpes
Behavior: Carnivore/
Nocturnal

Fennec Fox
◁ *Vulpes zerda*

THE FENNEC FOX SKULL is much like the European red fox at a glance, but closer investigation reveals larger orbits indicating better night vision, a lighter build to its skull and teeth, and a relatively larger cranium. Most pronounced, however, are the auditory bullae—as one might expect from an animal with such extraordinary ears. It is also the smallest species in the dog family.

Kingdom: Animalia
Phylum: Chordata
Class: Mammalia
Order: Carnivora

Family: Canidae
Genus: Vulpes
Behavior: Carnivore/
Nocturnal

Raccoon Dog
Nyctereutes procyonoides ◁

THE RACCOON DOG'S SKULL is heavily built and slightly elongated, with narrow zygomatic arches and a higher cranium than that of the raccoon. In older animals there is also a prominent sagittal crest. Raccoon dogs have flat molars and small, weak canines and carnassials, reflecting their omnivorous diet. They are primitive canines and not closely related to true raccoons.

AKA: Magnut
Kingdom: Animalia
Phylum: Chordata
Class: Mammalia
Order: Carnivora
Family: Canidae

Genus: Nyctereutes
Behavior: Omnivore/Nocturnal

Mammals
■ Carnivores

Gray Wolf
Canis lupus ▷

THIS GRAY WOLF skull is displayed gaping to show off its fearsome teeth. It possesses an elegant sagittal crest to support powerful jaws that can crush bone to extract nutritious marrow.

AKA: Wolf
Kingdom: Animalia
Phylum: Chordata
Class: Mammalia
Order: Carnivora

Family: Canidae
Genus: Canis
Behavior: Carnivore/
Nocturnal

Great Dane
Canis lupus familiaris ◁

THIS DOG SKULL BEARS a close resemblance to that of its gray wolf cousin. It has a distinct sagittal crest and a higher forehead and longer muzzle than those of other dogs.

Kingdom: Animalia
Phylum: Chordata
Class: Mammalia
Order: Carnivora

Family: Canidae
Genus: Canis
Behavior: Carnivore/
Diurnal

Pekingese Dog
Canis lupus familiaris

PEKINGESE DOGS are one of the most ancient varieties of toy dog. They were bred in China for their similarity to the "guardian lions" often seen outside important buildings. Their skull has a greatly reduced rostrum, resulting in their characteristic flat face.

AKA: Peke
Kingdom: Animalia
Phylum: Chordata
Class: Mammalia
Order: Carnivora
Family: Canidae
Genus: Canis
Behavior: Carnivore

Rottweiler
Canis lupus familiaris

THE ROTTWEILER'S SKULL has a distinctive high forehead and pronounced sagittal crest. The rottweiler is a very old breed, originally used for herding and droving domestic animals.

Kingdom: Animalia
Phylum: Chordata
Class: Mammalia
Order: Carnivora
Family: Canidae
Genus: Canis
Behavior: Carnivore/Diurnal

Spotted Hyena
Crocuta crocuta

OF ALL THE PREDATORS' skulls, that of the spotted hyena is surely the most robust and powerful. Although hyenas do hunt in packs, they also spend a lot of time scavenging and those immensely powerful jaws are capable of crushing the thickest of bones to reach the nutritious marrow within. Needless to say, there is a pronounced sagittal crest to which powerful muscles to attach.

AKA: Laughing Hyena   Family: Hyaenidae
Kingdom: Animalia   Genus: Crocuta
Phylum: Chordata   Behavior: Carnivore/
Class: Mammalia   Nocturnal
Order: Carnivora

DUDLEY'S NOTES:
You can see that the hyena's jaws are, indeed, the most powerful in the animal kingdom. I acquired this skull from a taxidermist I had to badger for nearly two years before he finally gave it to me. I make sure to display it with its mouth open.

Aardwolf
Proteles cristata

AS THE AARDWOLF'S appearance suggests, it is closely related to the hyena. However, the diet of this shy and nocturnal mammal is completely different: it eats almost exclusively termites, insect larvae, and occasionally carrion. Older aardwolves can lose some of their teeth, but this has little impact on their lifestyle.

AKA: Maanhaar Jackal
Kingdom: Animalia
Phylum: Chordata
Class: Mammalia
Order: Carnivora
Family: Hyaenidae
Genus: Proteles
Behavior: Insectivore/Nocturnal

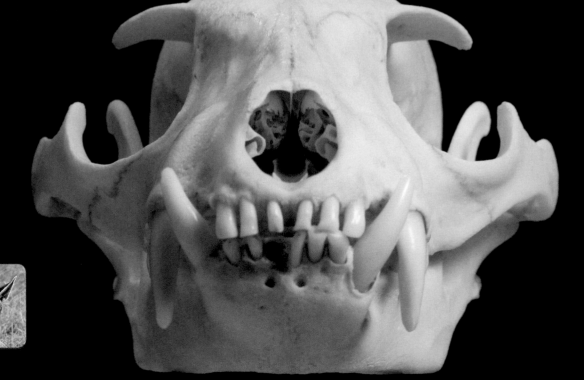

Southern Sea Lion
Otaria flavescens

THIS IS ONE OF the largest sea lions, found in the Antarctic region. There is enormous sexual dimorphism in the southern sea lion: males often grow to twice the size of the females. This almost bear-like skull appears to be one that has been exposed to the elements for some time.

AKA: Patagonian Sea Lion
Kingdom: Animalia
Phylum: Chordata
Class: Mammalia

Order: Carnivora
Family: Otariidae
Genus: Otaria
Behavior: Carnivore/ Aquatic

California Sea Lion
Zalophus californianus ◁

THIS IS THE CLASSIC circus seal: you know, the one that spins a ball on its head, claps, and blows a horn. But, strangely, it's not a true seal at all. The California sea lion is highly intelligent, and its pointed muzzle makes it look quite doglike. But strip them to the bone, and the differences are revealed—the sea lion has a more robust skull with teeth adapted as much for crushing as slicing. At sexual maturity, male California sea lions grow a large crest of bone on top of their heads.

Kingdom: Animalia
Phylum: Chordata
Class: Mammalia
Order: Carnivora

Family: Otariidae
Genus: Zalophus
Behavior: Piscivore/
Aquatic

Cape Fur Seal
Arctocephalus pusillus ◁

THE CAPE FUR SEAL is a dog like pinniped with a long snout. This seal closely resembles the California sea lion, but its distribution covers southern Africa and the waters around Australia, where its chief predator is the great white shark. Bleached skulls may be found in the deserts bordering the west coasts of South Africa and Namibia.

AKA: Brown Fur Seal,
South African Fur Seal
Kingdom: Animalia
Phylum: Chordata
Class: Mammalia
Order: Carnivora

Family: Otariidae
Genus: Arctocephalus
Behavior: Piscivore/
Aquatic

South American Fur Seal
Arctocephalus australis ▷

FUR SEAL SKULLS are very difficult to distinguish; this South American fur seal is very similar to its South African relation. There are nine species of fur seal worldwide. This skull reveals the powerful jaws that are equipped to tackle large crustaceans as well as fish.

Kingdom: Animalia
Phylum: Chordata
Class: Mammalia
Order: Carnivora

Family: Otariidae
Genus: Arctocephalus
Behavior: Piscivore/
Aquatic

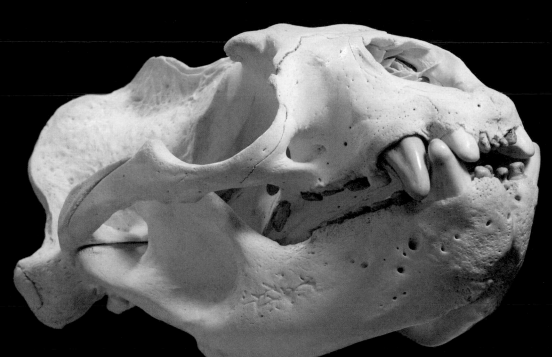

Steller Sea Lion
Eumetopias jubatus

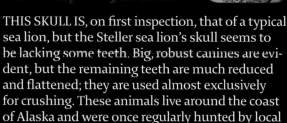

THIS SKULL IS, on first inspection, that of a typical sea lion, but the Steller sea lion's skull seems to be lacking some teeth. Big, robust canines are evident, but the remaining teeth are much reduced and flattened; they are used almost exclusively for crushing. These animals live around the coast of Alaska and were once regularly hunted by local Inuit communities.

AKA: Northern Sea Lion
Kingdom: Animalia
Phylum: Chordata
Class: Mammalia
Order: Carnivora

Family: Otariidae
Genus: Eumetopias
Behavior: Piscivore/ Aquatic

Subantarctic Fur Seal
Arctocephalus tropicalis

COMPARED TO THOSE of other seals, the subantarctic fur seal's skull is blunt, and there is little evidence of the powerful muscle attachment to the rear of the skull. It has a sharp set of teeth, used predominantly for catching small fish and squid.

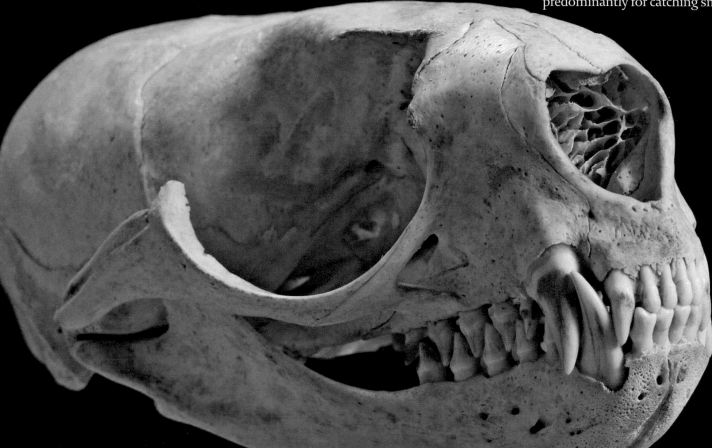

Kingdom: Animalia
Phylum: Chordata
Class: Mammalia
Order: Carnivora
Family: Otariidae
Genus: Arctocephalus
Behavior: Piscivore/ Aquatic

Hooded Seal
Cystophora cristata ▷

THE MOST OBVIOUS FEATURE of a hooded seal is not at all evident in its skull. At about four years of age, a bulge develops on top of the male's nose that can be inflated during courtship displays. This Arctic-dwelling seal has the shortest nursing period of any mammal: the pups get to drink their mother's rich milk for an average of just four days.

Kingdom: Animalia
Phylum: Chordata
Class: Mammalia
Order: Carnivora

Family: Phocidae
Genus: Cystophora
Behavior: Piscivore/ Aquatic

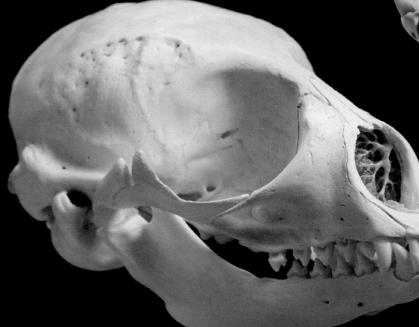

Harp Seal
Pagophilus groenlandicus ◁

THIS IS THE MOST delicate seal skull in the collection. The harp seal is a lover of cold seas; it breeds on pack ice in the Arctic. It is also strongly migratory, traveling thousands of miles in search of food.

AKA: Saddleback Seal
Kingdom: Animalia
Phylum: Chordata
Class: Mammalia

Order: Carnivora
Family: Phocidae
Genus: Pagophilus
Behavior: Piscivore/ Aquatic

Harbor Seal
Phoca vitulina ▷

THIS IS AN ELEGANT, elongated carnivore skull with small canine teeth and a row of sharp carnassials. Harbor seals are a widely distributed pinnipeds, found around the temperate and Arctic coasts of the Northern Hemisphere.

AKA: Common Seal
Kingdom: Animalia
Phylum: Chordata
Class: Mammalia
Order: Carnivora

Family: Phocidae
Genus: Phoca
Behavior: Piscivore/ Aquatic

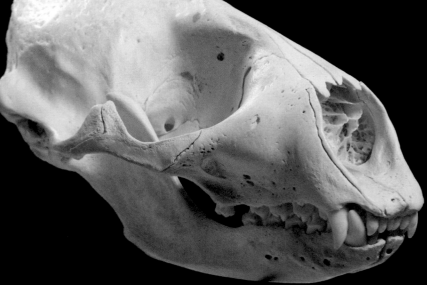

Meerkat

Suricata suricatta ▷

THIS SKULL IS UNUSUAL in that its orbits are squarish. Meerkats are burrowing animals that live in highly social colonies in the arid regions of southern Africa. They are fond of munching on invertebrates and have developed immunity to scorpion venom. Meerkats are well known for their characteristically vigilant upright stance.

AKA: Suricate
Kingdom: Animalia
Phylum: Chordata
Class: Mammalia
Order: Carnivora
Family: Herpestidae
Genus: Suricata
Behavior:
Insectivore/
Diurnal

Binturong

Arctictis binturong ◁

RELATED TO CIVETS and genets, this arboreal mammal is found in the dense rainforests of Southeast Asia. Its poor eyesight and well-developed sense of smell are reflected in its skull morphology.

AKA: Bearcat
Kingdom: Animalia
Phylum: Chordata
Class: Mammalia
Order: Carnivora
Family: Viverridae
Genus: Arctictis
Behavior: Omnivore/
Nocturnal

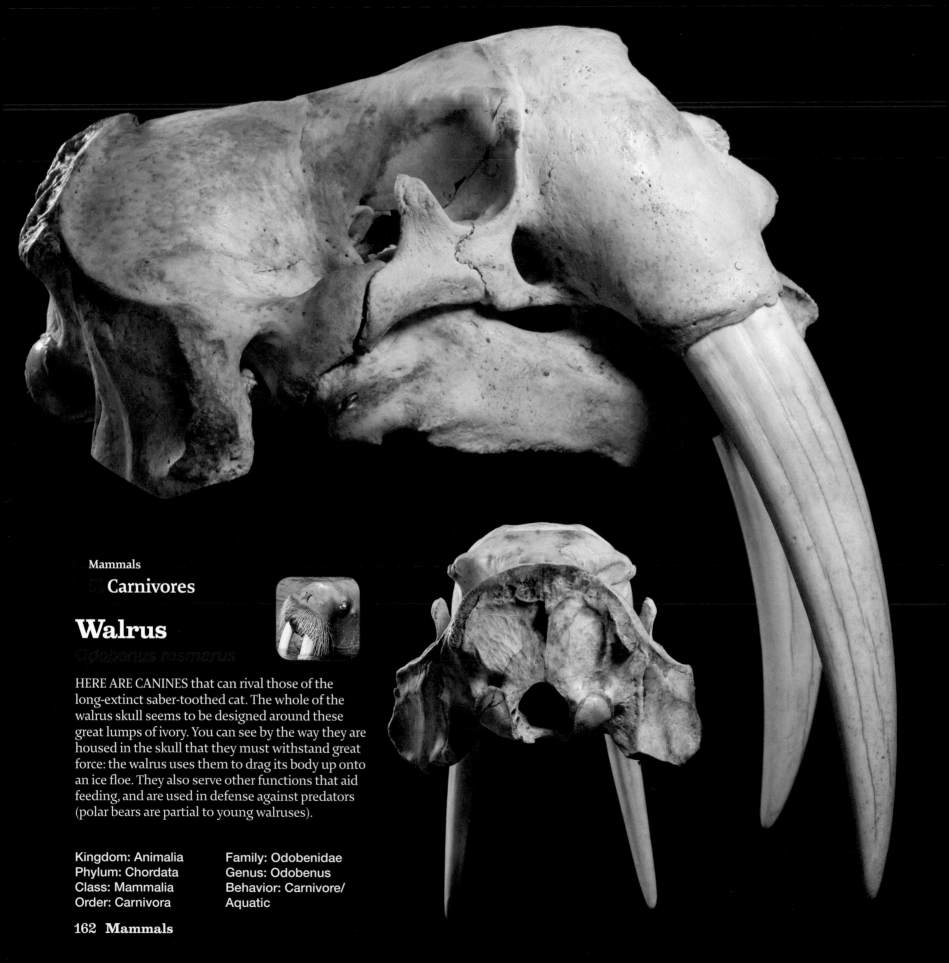

Mammals
Carnivores

Walrus
Odobenus rosmarus

HERE ARE CANINES that can rival those of the
long-extinct saber-toothed cat. The whole of the
walrus skull seems to be designed around these
great lumps of ivory. You can see by the way they are
housed in the skull that they must withstand great
force: the walrus uses them to drag its body up onto
an ice floe. They also serve other functions that aid
feeding, and are used in defense against predators
(polar bears are partial to young walruses).

Kingdom: Animalia
Phylum: Chordata
Class: Mammalia
Order: Carnivora

Family: Odobenidae
Genus: Odobenus
Behavior: Carnivore/
Aquatic

Raccoon
▷ *Procyon lotor*

RACCOON SKULLS ARE SMALL with a broad, smooth, and rounded cranium and no obvious sagittal crest. The rostrum is long and pointed, and the auditory bullae are rounded and inflated on one side. The most distinctive feature of the raccoon skull it is its palate, which extends far past the last molar tooth (sadly not visible here). The orbits are large and forward-facing. Its teeth reflect its omnivorous diet: broad, flat-topped molars, incisors for nipping, and pointed canines.

AKA: North American
Raccoon
Kingdom: Animalia
Phylum: Chordata
Class: Mammalia

Order: Carnivora
Family: Procyonidae
Genus: Procyon
Behavior: Omnivore/
Nocturnal

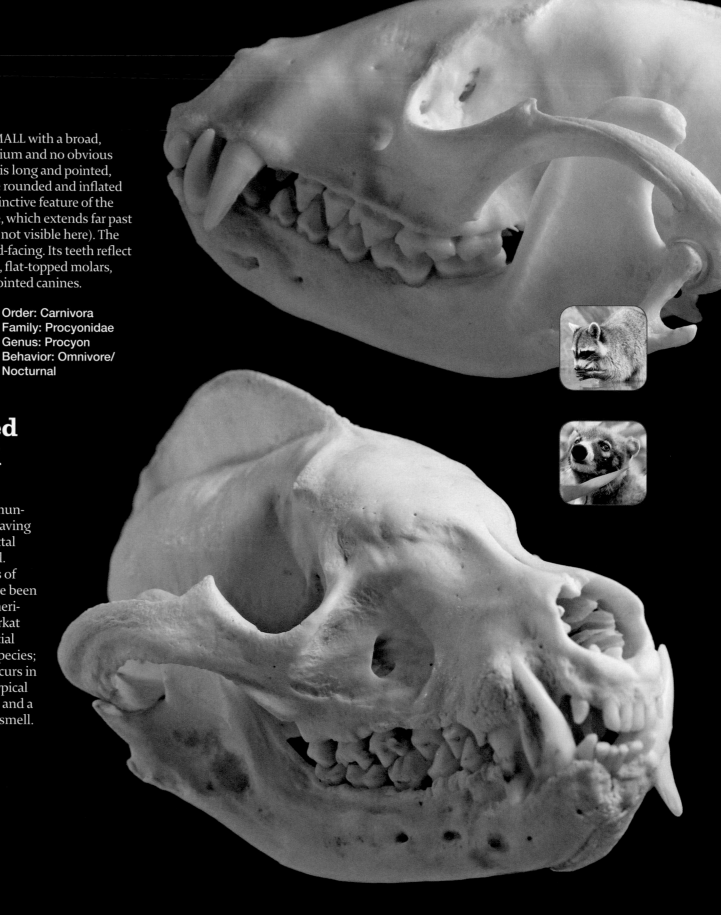

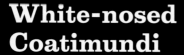

White-nosed Coatimundi
▷ *Nasua narica*

THE WHITE-NOSED coatimundi's skull is distinctive in having a large, almost finlike sagittal crest at the back of its skull. Coatimundis are members of the raccoon family but have been described as the South American equivalent of the meerkat because it lives in large social groups. There are several species; the white-nosed variety occurs in Central America. It has a typical skull of a small insectivore and a highly developed sense of smell.

AKA: Ring-tailed
Coatimundi
Kingdom: Animalia
Phylum: Chordata
Class: Mammalia
Order: Carnivora
Family: Procyonidae
Genus: Nasua
Behavior: Omnivore/
Diurnal

Kinkajou
△ *Potos flavus*

THE KINKAJOU'S omnivorous diet consists of a high proportion of fruit. The skull exhibits the large orbits of a strictly nocturnal animal. This animal is found in Central and South America and has a particular fondness for figs.

AKA: Honey Bear
Kingdom: Animalia
Phylum: Chordata
Class: Mammalia
Order: Carnivora
Family: Procyonidae
Genus: Potos
Behavior: Omnivore/ Nocturnal

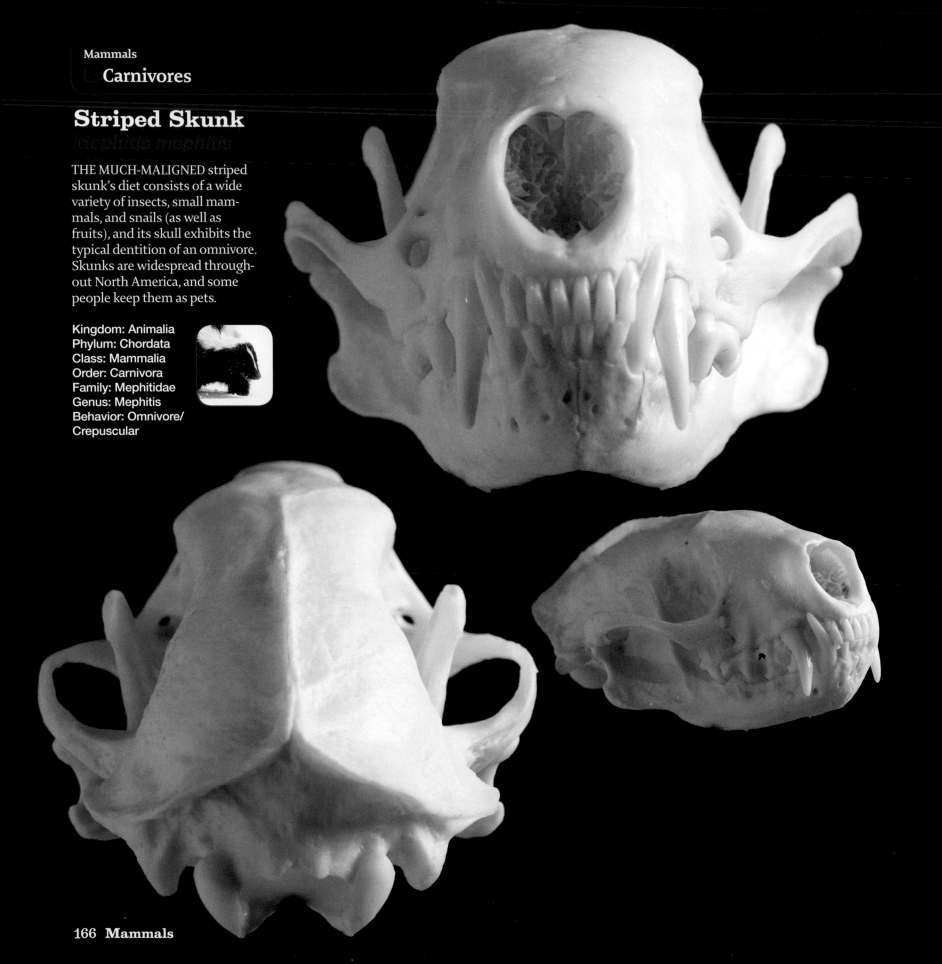

Striped Skunk
Mephitis mephitis

THE MUCH-MALIGNED striped skunk's diet consists of a wide variety of insects, small mammals, and snails (as well as fruits), and its skull exhibits the typical dentition of an omnivore. Skunks are widespread throughout North America, and some people keep them as pets.

Kingdom: Animalia
Phylum: Chordata
Class: Mammalia
Order: Carnivora
Family: Mephitidae
Genus: Mephitis
Behavior: Omnivore/
Crepuscular

American Badger

Taxidea taxus ◁

THIS SKULL IS similar to that of the European badger, but the American badger has a more carnivorous diet—reflected in its impressive teeth and powerful jaw.

Kingdom: Animalia
Phylum: Chordata
Class: Mammalia
Order: Carnivora

Family: Mustelidae
Genus: Taxidea
Behavior: Carnivore/
Nocturnal

Hog Badger

Arctonyx collaris ▷

THERE ARE THREE TYPES of badger skulls in the collection. This one hails from Southeast Asia and, although smaller, is similar to the skull of the European badger. Badgers are omnivorous creatures with a fondness for worms.

Kingdom: Animalia
Phylum: Chordata
Class: Mammalia
Order: Carnivora
Family: Mustelidae
Genus: Arctonyx
Behavior: Carnivore/
Nocturnal

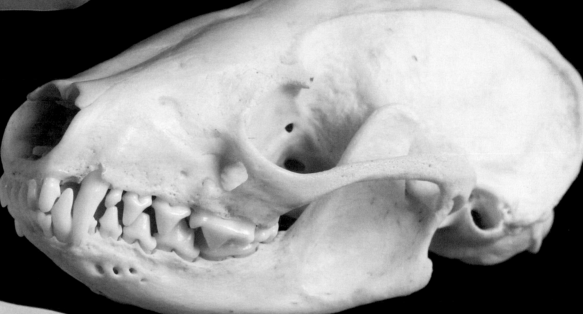

European Badger

Meles meles ◁

THIS SKULL HAS all the classic attributes of a carnivore—incisors, carnassials, and pronounced sagittal crest. Its relatively small orbits and elongated snout indicate that it hunts more by smell than sight.

Kingdom: Animalia
Phylum: Chordata
Class: Mammalia
Order: Carnivora
Family: Mustelidae
Genus: Meles
Behavior: Omnivore/Nocturnal

Asian Small-clawed Otter
Amblonyx cinereus

THIS IS THE SMALLEST of all otter species. It has a flattened and streamlined head with eyes towards the front. Otters feed mainly on small crustaceans. They often hunt underwater and rely primarily on touch and smell to locate prey. Their teeth are therefore broad and robust to enable them to crush the shells of crustaceans. Due to widespread habitat destruction in Southeast Asia, this otter is now on the IUCN Red List of Threatened Species.

AKA: Oriental Small-clawed Otter
Kingdom: Animalia
Phylum: Chordata
Class: Mammalia

Order: Carnivora
Family: Mustelidae
Genus: Amblonyx
Behavior: Carnivore/ Diurnal

European Otter
Lutra lutra

OTTERS HAVE A SMALL, flattened, and elongated skull that is streamlined to suit their aquatic lifestyle. They are largely nocturnal animals, and as such, their orbits are small; smell plays a much more important sensory role than sight. The skull is unmistakably that of a carnivore.

AKA: Old World Otter
Kingdom: Animalia
Phylum: Chordata
Class: Mammalia

Order: Carnivora
Family: Mustelidae
Genus: Lutra
Behavior: Piscivore/ Diurnal

Fisher
Martes pennanti

THE FISHER SKULL is most notable for its sexual dimorphism. Although otherwise typical of small martens, weasels, and ferrets, the skull of the male fisher has a fine sagittal crest, almost like a sail at the back of the head. Usually for muscle attachment, in this case the crest is also an indicator of sexual maturity.

AKA: Fisher Cat
Kingdom: Animalia
Phylum: Chordata
Class: Mammalia
Order: Carnivora
Family: Mustelidae
Genus: Martes
Behavior: Carnivore/
Nocturnal

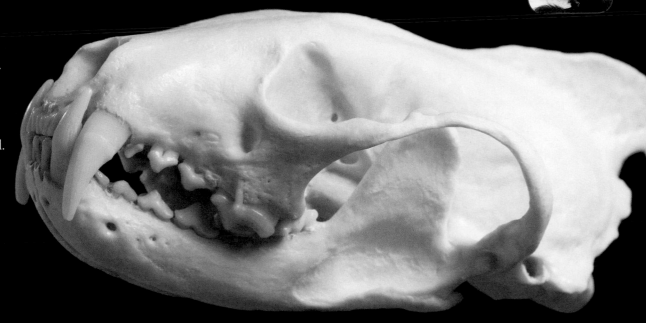

Least Weasel
Mustela nivalis

THIS SMALL, ELONGATED, least weasel skull seems separated into two halves—the front part for the sense organs, the rear for the brain. Notice also the delicately thin zygomatic arches and the dagger-like canines.

AKA: Weasel
Kingdom: Animalia
Phylum: Chordata
Class: Mammalia
Order: Carnivora

Family: Mustelidae
Genus: Mustela
Behavior: Carnivore/
Crepuscular

Wolverine
Gulo gulo

THE WOLVERINE is the largest terrestrial member of the weasel family and is found in northern boreal forests. Its skull is robust with wide zygomatic arches. Wolverines are notoriously ferocious predators and have even been known to attack adult deer. They grow to the size of an average dog and have dark, oily fur that helps to prevent frost and snow settling into it.

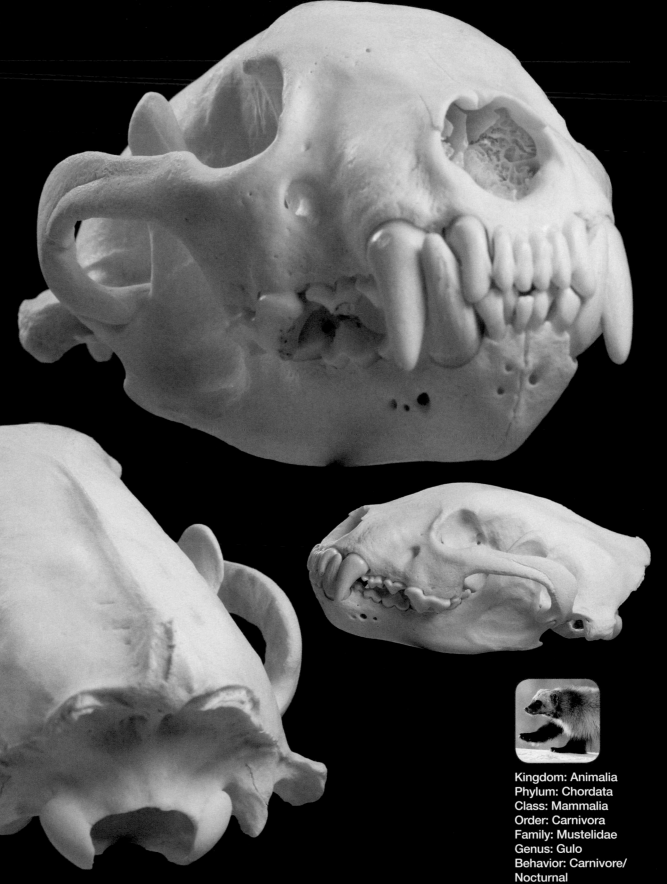

Kingdom: Animalia
Phylum: Chordata
Class: Mammalia
Order: Carnivora
Family: Mustelidae
Genus: Gulo
Behavior: Carnivore/ Nocturnal

Duck-billed Platypus

Ornithorhynchus anatinus

IS IT A BIRD? Is it a reptile? Or is it a mammal? In fact it's a duck-billed platypus, a primitive egg-laying member of the mammal family. Like all true mammals, they possess a malleus, incus, and stapes—but in the platypus these three bones remain large and joined to the rest of the skull rather than floating in the middle ear to conduct sound from the eardrum. Another ancestral skull feature is the external opening of the ear at the base of the jaw (in most other mammals it opens above). Young animals possess three-cusped molar teeth, which they lose as their "beak" develops and hardens.

AKA: Platypus
Kingdom: Animalia
Phylum: Chordata
Class: Mammalia
Order: Monotremata

Family: Ornithorhynchidae
Genus: Ornithorhynchus
Behavior: Insectivore/ Nocturnal

Pronghorn
Antilocapra americana ▷

THE PRONGHORN'S unique bifurcating horns are distinctive; they are covered in a keratin sheath that is shed annually. The orbits sit high on the skull just below the horns. Pronghorns are notable for their speed and are often cited as the second-fastest land mammals after the cheetah.

AKA: Pronghorn Antelope
Kingdom: Animalia
Phylum: Chordata
Class: Mammalia
Order: Artiodactyla
Family: Antilocapridae
Genus: Antilocapra
Behavior: Herbivore/ Crepuscular

Giraffe
Giraffa camelopardalis ▽

ADULT GIRAFFES HAVE distinct, hair-covered horns called ossicones. Males use these horns to fight with one another. As male giraffes grow older, calcium deposits form on their skulls, and another horn-like bump develops in the center of the skull. These horns are absent on the skull of the newborn giraffe. The adult giraffe skull is very large with a long diastema, which accommodates an extendable, prehensile tongue.

Kingdom: Animalia
Phylum: Chordata
Class: Mammalia
Order: Artiodactyla
Family: Giraffidae
Genus: Giraffa
Behavior: Herbivore/ Diurnal

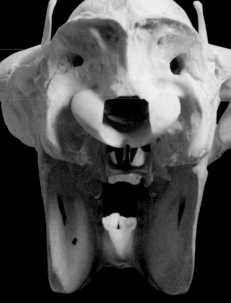

Alpaca
Vicugna pacos ◁

THE ALPACA SKULL resembles that of a small llama. Note the extension of its jawbone towards the top of the head—almost like horns. The alpaca is a native of South America, now entirely domesticated and prized for its warm and silky wool.

Kingdom: Animalia
Phylum: Chordata
Class: Mammalia
Order: Artiodactyla

Family: Camelidae
Genus: Vicugna
Behavior: Herbivore/
Diurnal

Bactrian Camel
Camelus bactrianus ▷

THE BACTRIAN is the two-humped variety of camel. More than a million of these camels have been domesticated, but there are fewer than 800 roaming the wilds of Mongolia, giving it Critically Endangered status on the IUCN Red List. It has a shorter rostrum than its single-humped cousin but a longer cranium.

Kingdom: Animalia
Phylum: Chordata
Class: Mammalia
Order: Artiodactyla

Family: Camelidae
Genus: Camelus
Behavior: Herbivore/
Diurnal

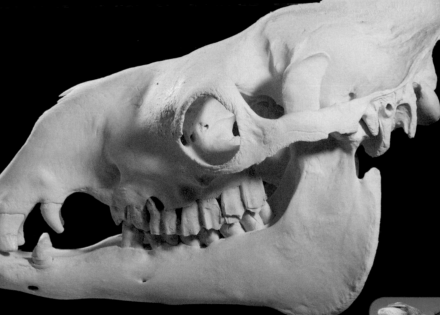

Dromedary Camel
Camelus dromedarius ◁

THE DROMEDARY is the more familiar form of camel, with one hump. The skull is similar to that of its two-humped bactrian camel cousin and clearly that of a herbivore with incisors that meet a flat palate in the upper jaw. It is now domesticated throughout the Middle East; the only place it demonstrates wild behavior is in the outback of Australia, where there is a burgeoning feral population.

AKA: Arabian Camel
Kingdom: Animalia
Phylum: Chordata
Class: Mammalia
Order: Artiodactyla

Family: Camelidae
Genus: Camelus
Behavior: Herbivore/
Diurnal

Muntjac
Muntiacus muntjak △

WHILE MALE MUNTJACS often grow small and slender antlers like those of other deer, they tend to use their curious tusks (a pair of downward-pointing canine teeth) as weapons. The antlers join a ridge that extends down beyond the eyes to give the skull an elegant, streamlined appearance.

AKA: Barking Deer
Kingdom: Animalia
Phylum: Chordata
Class: Mammalia
Order: Artiodactyla
Family: Cervidae
Genus: Muntiacus
Behavior: Omnivore/
Crepuscular

Roe Deer
▷ *Capreolus capreolus*

ONLY MALE ROE DEER grow antlers, which are shed each year and get larger in subsequent years, growing up to three or even four points. This is a typical herbivore skull with rows of molars separated from a set of lower incisors by a diastema. The incisors press against a bony palate in the upper jaw.

AKA: Western Roe Deer
Kingdom: Animalia
Phylum: Chordata
Class: Mammalia
Order: Artiodactyla
Family: Cervidae
Genus: Capreolus
Behavior: Herbivore/
Crepuscular

AKA: Vampire Deer
Kingdom: Animalia
Phylum: Chordata
Class: Mammalia
Order: Artiodactyla
Family: Cervidae
Genus: Hydropotes
Behavior: Herbivore/
Diurnal

Chinese Water Deer

△ *Hydropotes inermis*

THIS UNUSUAL SPECIES of deer doesn't grow antlers—instead it
has tusks (actually, massively enlarged upper canine teeth; hence the
name vampire deer in North America). These impressive tusks can
grow as long as eight centimeters (three inches) in the bucks and
can also be used to age an animal. They are held only loosely in their
sockets and can be "wiggled" and even moved out of the way while
eating, but are displayed forward during the rut. Feral populations of
Chinese water deer now live in parts of Europe and North America.

■ Even-toed Ungulates

Hippopotamus
Hippopotamus amphibius

OF THE LAND MAMMALS, the skull of the hippo sits alongside those of the Javan rhino and African elephant as among the most impressive. This skull is the biggest in Dudley's collection; its huge ivory teeth indicate that it came from a particularly large male hippo. The orbits seem to enclose the eye socket and rise above the rest of the skull. This allows the hippo to see while wallowing almost submerged in water—much like the Nile crocodile with which it shares its habitat.

Kingdom: Animalia
Phylum: Chordata
Class: Mammalia
Order: Artiodactyla
Family: Hippootamidae
Genus: Hippopotamus
Behavior: Herbivore/ Diurnal

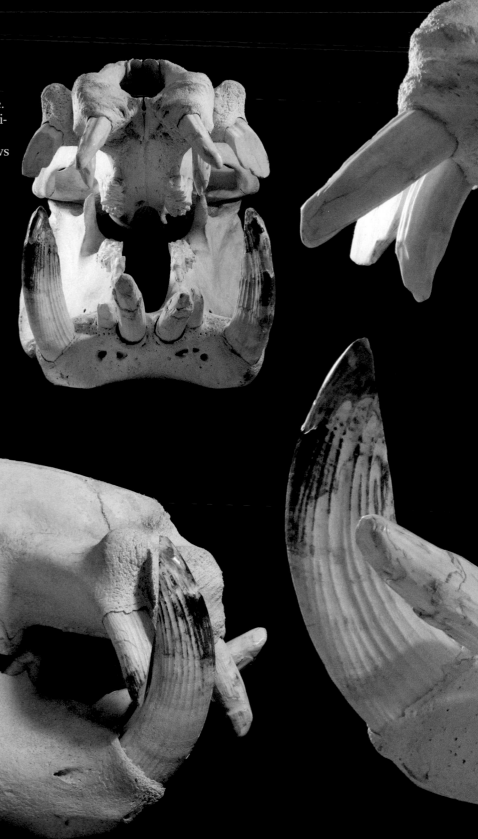

Even-toed Ungulates

North Sulawesi Babirusa

Babyrousa babyrussa

THE MALE BABIRUSA has highly conspicuous tusks formed from the upper canine teeth, which grow upwards, almost like antlers; their exact shape depends on the species. In the North Sulawesi babirusa, they grow upwards through the skull and curve back towards the skull between the eyes; in the Togian babirusa, they lack the strong curve. In both species, the lower canines grow large, much like those of a warthog.

Kingdom: Animalia
Phylum: Chordata
Class: Mammalia
Order: Artiodactyla
Family: Suidae
Genus: Babyrousa
Behavior: Omnivore/
Nocturnal

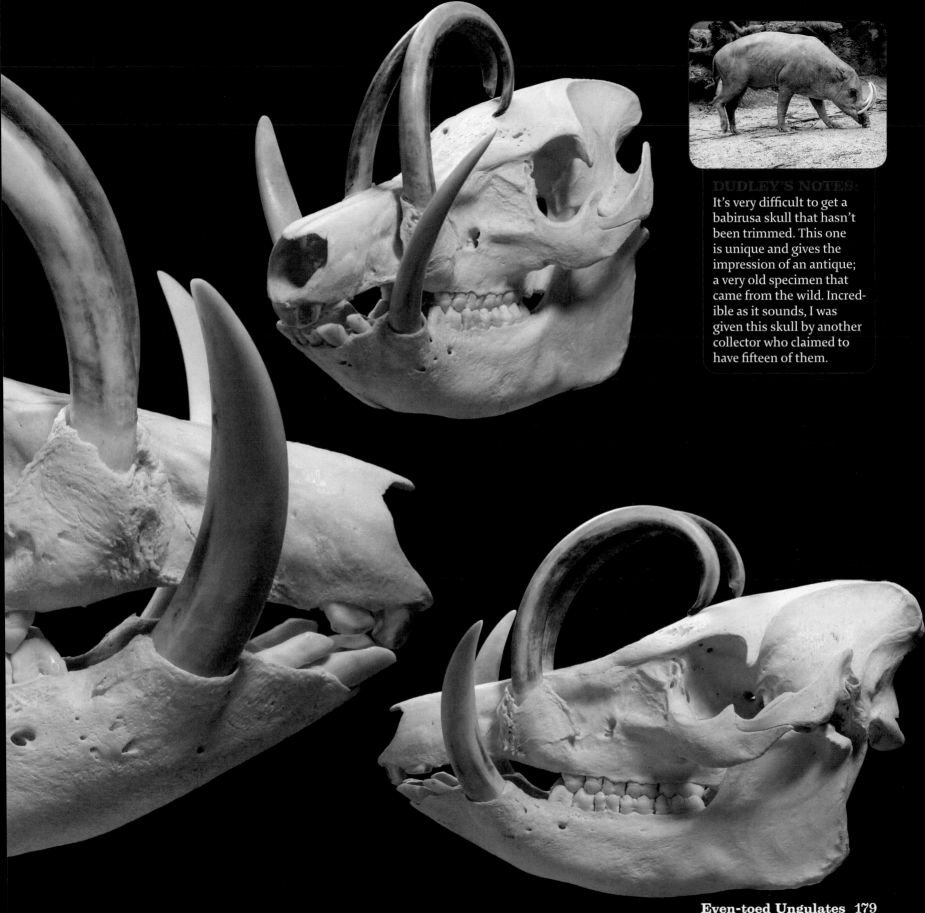

Domestic Pig
Sus domestica ▷

WITHOUT CAREFUL TRIMMING, the tusks of domestic pigs can reach fearsome lengths. To add to the danger, the top tusks rub against the bottom tusks, sharpening them as they grow.

Kingdom: Animalia
Phylum: Chordata
Class: Mammalia
Order: Artiodactyla
Family: Suidae

Genus: Sus
Behavior:Herbivore/
Diurnal

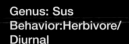

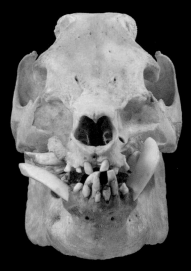

Vietnamese Pot-bellied Pig
Sus domestica ▷

THOUSANDS OF YEARS of selective breeding accounts for the shortened snout and curious teeth of this pig skull. The snout is shortened and the brow raised, resulting in a steep and almost concave profile. Vietnamese pot-bellied pigs are smaller than most other domestic breeds and originate in mountainous areas of Vietnam and Thailand.

Kingdom: Animalia
Phylum: Chordata
Class: Mammalia
Order: Artiodactyla
Family: Suidae
Genus: Sus
Behavior: Omnivore/
Diurnal

African Bushpig
Potamochoerus larvatus ▷

ALTHOUGH CANINE TUSKS are clearly evident in this skull, the African bushpig doesn't grow the huge curling tusks that the closely related warthog does. As with other types of pig, the primary sense is that of smell, and this is reflected in the skull architecture: small eye sockets (in this case high on the skull) and a large, well-developed nasal region.

AKA: Savannah Bushpig
Kingdom: Animalia
Phylum: Chordata
Class: Mammalia
Order: Artiodactyla
Family: Suidae
Genus: Potamochoerus
Behavior: Herbivore/ Diurnal

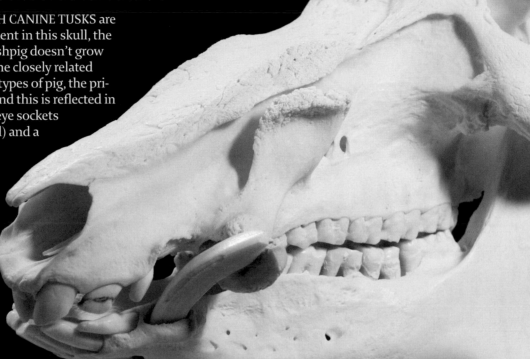

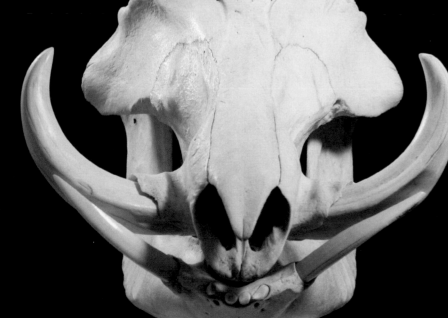

Warthog
Phacochoerus africanus ◁

AS MEMBERS OF THE pig family go, this is an elegant skull. Two pairs of sharp tusks protrude out either side of the mouth. These are self-sharpening against an upper tooth and have been known to inflict serious damage on predators and incautious hunters.

AKA: Common Warthog
Kingdom: Animalia
Phylum: Chordata
Class: Mammalia
Order: Artiodactyla
Family: Suidae
Genus: Phacochoerus
Behavior: Omnivore/Diurnal

> **DUDLEY'S NOTES:**
> The warthog is not a particularly attractive animal. The skull, however, is incredible. It's obviously a pig, but it's got amazing tusks. This one is a good example, but I've seen photographs of a warthog skull in which the tusks have grown in a full circle.

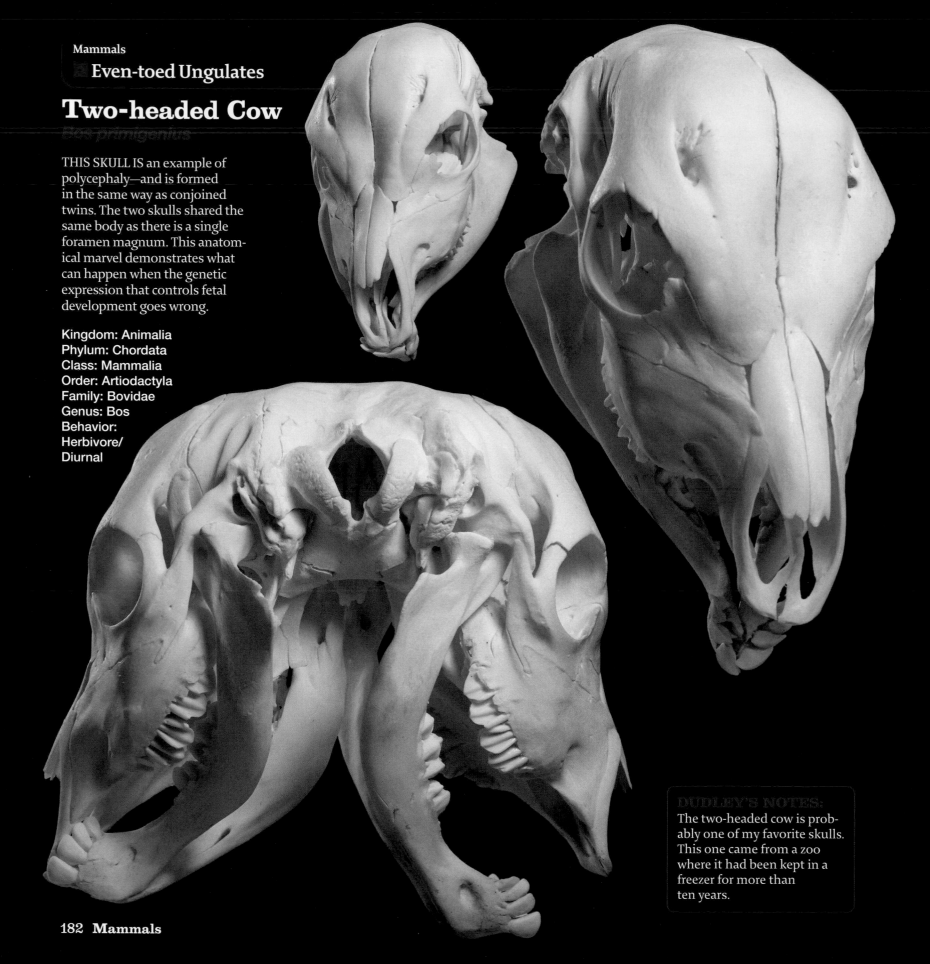

Two-headed Cow
Bos primigenius

THIS SKULL IS an example of polycephaly—and is formed in the same way as conjoined twins. The two skulls shared the same body as there is a single foramen magnum. This anatomical marvel demonstrates what can happen when the genetic expression that controls fetal development goes wrong.

Kingdom: Animalia
Phylum: Chordata
Class: Mammalia
Order: Artiodactyla
Family: Bovidae
Genus: Bos
Behavior:
Herbivore/
Diurnal

DUDLEY'S NOTES:
The two-headed cow is probably one of my favorite skulls. This one came from a zoo where it had been kept in a freezer for more than ten years.

▪ Even-toed Ungulates

Black Wildebeest
Connochaetes gnou

THIS IMPRESSIVELY HORNED black wildebeest skull is displayed without its mandible.
Wildebeests are a type of antelope, and the
skull has the broad flat palate of a grazing
animal. This type of wildebeest is found
in South Africa.

AKA: White-tailed Gnu
Kingdom: Animalia
Phylum: Chordata
Class: Mammalia
Order: Artiodactyla
Family: Bovidae
Genus: Connochaetes
Behavior: Herbivore/
Diurnal

Blackbuck
Antilope cervicapra

THIS IS AN INDIAN ANTELOPE, which can be found on the IUCN Red List. Males and females are very different in color, and the females also lack horns. It is also notable as one of the fastest of all antelopes (perhaps because historically its chief predator was the now extinct Indian cheetah). Blackbuck are now raised on ranches in North America, where they are hunted for their "trophy" horns.

Kingdom: Animalia
Phylum: Chordata
Class: Mammalia
Order: Artiodactyla

Family: Bovidae
Genus: Antilope
Behavior: Herbivore/
Diurnal

American Blackbelly
Ovis aries

THIS AMERICAN BLACKBELLY ram skull earns its place in the collection due to a particularly impressive set of horns that complete a full rotation. The American blackbelly is a result of crossing the mouflon with the Barbados blackbelly, to produce a distinctively marked sheep with large horns.

AKA: Sheep
Kingdom: Animalia
Phylum: Chordata
Class: Mammalia
Order: Artiodactyla

Family: Bovidae
Genus: Ovis
Behavior: Herbivore/
Diurnal

American Bison
Bison bison ◁

THE AMERICAN BISON is very similar to the European bison, but it has smaller horns and a slightly shorter rostrum, probably because it tends to fight by charging rather than locking horns. This animal and its skull were once symbols of power among Native Americans. Famously, the American bison was hunted to near-extinction by European colonists.

Kingdom: Animalia   Family: Bovidae
Phylum: Chordata   Genus: Bison
Class: Mammalia   Behavior: Herbivore/
Order: Artiodactyla   Diurnal

Domestic Sheep
Ovis aries ▷

THE RAM'S HORN has long been used in religious rites: the shofar is frequently mentioned in the Bible and the bukkehorn in Norse mythology. Like the shofar, the bukkehorn is an ancient musical instrument. It was traditionally used by Norwegian cowherds on summer dairy farms in the high mountains. Dudley's specimen has unimpressive horns.

AKA: Ram
Kingdom: Animalia
Phylum: Chordata
Class: Mammalia
Order: Artiodactyla
Family: Bovidae
Genus: Ovis
Behavior: Herbivore/
Diurnal

Forest Buffalo
Syncerus caffer nanus ◁

THIS SKULL IS MISSING its mandible, so it looks like a small version of the type of skull one sometimes sees mounted on the front of big American cars or hung above a gate at a ranch entrance. The forest buffalo is a diminutive subspecies found in the forests of central and western Africa.

Kingdom: Animalia
Phylum: Chordata
Class: Mammalia
Order: Artiodactyla
Family: Bovidae
Genus: Syncerus
Behavior: Herbivore/Diurnal

Mouflon
Ovis orientalis

THE MOUFLON is a wild variety of sheep considered to be one of the ancestors of modern domestic sheep. The skull is notable for its fine horns that are used in head-butting contests among males. Recently, the mouflon has been introduced into the southern states of the U.S. for the entertainment of trophy hunters.

Kingdom: Animalia
Phylum: Chordata
Class: Mammalia
Order: Artiodactyla
Family: Bovidae
Genus: Ovis
Behavior: Herbivore/
Diurnal

Zebu
Bos primigenius

THE ZEBU IS an Asian cow with a distinctive pair of horns, a hump on its back, and a large dewlap. These are the animals that carelessly wander the roads of the Indian subcontinent. The zebu (there are estimated to be more than seventy breeds) is thought to be one of the more primitive varieties of cow, perhaps as old as the now-extinct aurochs.

Kingdom: Animalia
Phylum: Chordata
Class: Mammalia
Order: Artiodactyla
Family: Bovidae
Genus: Bos
Behavior: Herbivore/ Diurnal

Burchell's Zebra
Equus quagga burchelli ◁

THE BURCHELL'S ZEBRA is a member of the horse family and its skull shares many characteristics with that of the domestic horse. The zebra's tall cheek teeth need to survive a high degree of erosion from a grass diet that is rich in silica (basically, glass). A trained eye can easily determine the age of a zebra (or horse) by examining the wear on its teeth. Which is why we might "look a gift horse in the mouth."

Kingdom: Animalia
Phylum: Chordata
Class: Mammalia
Order: Perissodactyla

Family: Equidae
Genus: Equus
Behavior: Herbivore/ Diurnal

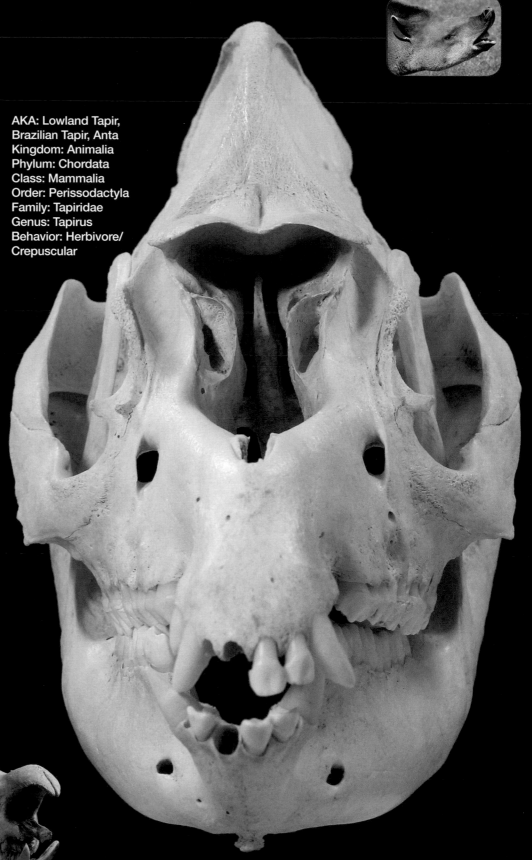

South American Tapir
▷ *Tapirus terrestris*

THIS IS A VERY distinctive skull with bony ridges above the eyes and a cranium that narrows anteriorly. The South American tapir is a large beast, growing to over 225 kilograms (500 pounds) and has long been hunted by the indigenous people of the Amazon Basin. It spends a lot of time in water and is a good swimmer, able to use its mobile snout as a snorkel. Worldwide, there are four species of tapir, three in South America and one in the jungles of Malaysia and Sumatra.

AKA: Lowland Tapir,
Brazilian Tapir, Anta
Kingdom: Animalia
Phylum: Chordata
Class: Mammalia
Order: Perissodactyla
Family: Tapiridae
Genus: Tapirus
Behavior: Herbivore/
Crepuscular

Javan Rhinoceros
◁ *Rhinoceros sondaicus*

THE JAVAN RHINOCEROS is one of the world's rarest mammals. It was declared extinct in Vietnam in October 2011 and is now found only in a few areas of dense, inaccessible jungle in Java. This skull is from the Oxford Museum of Natural History, and happened to be lying nearby when we photographed the dodo and the Piltdown Man cast. Its keratin horn is missing, but you can see the flattened area on the top of the rostrum where it once sat.

AKA: Lesser One-horned Rhinoceros
Kingdom: Animalia
Phylum: Chordata
Class: Mammalia
Order: Perissodactyla
Family: Rhinocerotidae
Genus: Rhinoceros
Behavior: Herbivore/Diurnal

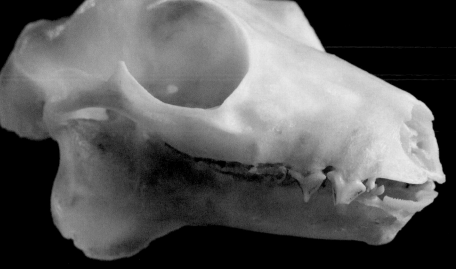

Philippine Flying Lemur
Cynocephalus volans ◁

NOT REALLY A LEMUR at all, this animal belongs to a genus all of its own. Its teeth are unusual; the lower incisors are flat and split and possess many comblike forks. There are no upper incisors, just a palate. This unique arrangement of teeth may be used for specialized feeding and grooming activities.

Kingdom: Animalia
Phylum: Chordata
Class: Mammalia
Order: Dermoptera

Family: Cynocephalidae
Genus: Cynocephalus
Behavior: Herbivore/
Nocturnal

European Hedgehog
Erinaceus europaeus ▷

THE EUROPEAN HEDGEHOG skull is easily recognized by a pair of prominent, blunt upper front teeth with a gap in between. The small, broad skull contains a relatively small brain. Hedgehogs are omnivorous and have a fondness for juicy invertebrates.

AKA: Common Hedgehog
Kingdom: Animalia
Phylum: Chordata
Class: Mammalia
Order: Erinaceomorpha
Family: Erinaceidae
Genus: Erinaceus
Behavior: Carnivore/Nocturnal

DUDLEY'S NOTES:
When I found this hedgehog it was in perfect condition. It was absolutely pristine with no bones or teeth missing. It would be wonderful if every skull I found were like this.

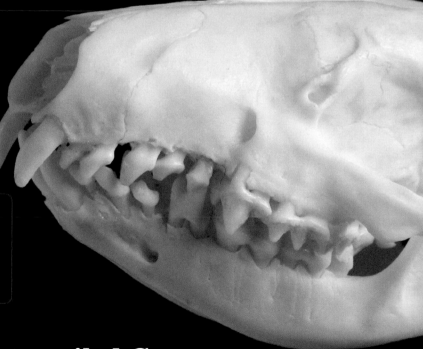

Short-tailed Gymnure
Hylomys suillus ◁

The short-tailed gymnure is a shrewlike creature, related to the hedgehogs, that inhabits the jungles of Eastern Borneo, Thailand, Sumatra, and Myanmar. The gymnure's rostrum is slightly longer than in hedgehogs; as a result, gymnures have a narrower, more shrew-like snout.

Kingdom: Animalia
Phylum: Chordata
Class: Mammalia
Order: Erinaceomorpha

Family: Erinaceidae
Genus: Hylomys
Behavior: Carnivore/
Nocturnal

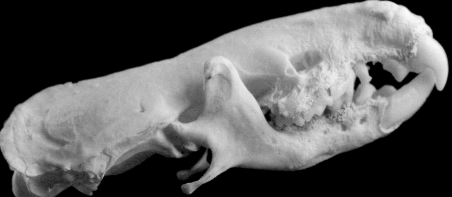

Common Spotted Cuscus
Spilocuscus maculatus ▷

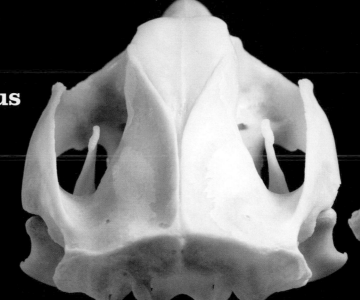

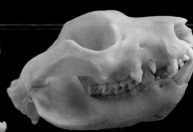

THIS SPOTTED CUSCUS SKULL, when viewed from the rear, has an elegant collection of ridges and grooves formed by the zygomatic arches. This animal is a marsupial and lives almost exclusively on the island of New Guinea. It is about the size of a cat and has a prehensile tail.

Kingdom: Animalia
Phylum: Chordata
Class: Mammalia
Order: Diprotodontia

Family: Phalangeroidea
Genus: Spilocuscus
Behavior: Herbivore/
Nocturnal

Red Kangaroo
Macropus rufus ◁

THE RED KANGAROO is the largest marsupial and is widespread in Australia. This large skull has its mandible hinged right at the back—unlike most placental mammals but typical of kangaroos. Kangaroos are prodigious jumpers and can leap a huge nine meters (thirty feet) in a single bound.

Kingdom: Animalia
Phylum: Chordata
Class: Mammalia
Order: Diprotodontia
Family: Macropodidae
Genus: Macropus
Behavior: Herbivore/Nocturnal

Long-nosed Potoroo
Potorous tridactylus ◁

THE SKULL OF THIS MARSUPIAL bears a close resemblance to that of the long-nosed bandicoot. In the flesh, the long-nosed potoroo looks like a rat or a shrew (and indeed occupies a similar ecological niche). However, it hops like a miniature kangaroo.

Kingdom: Animalia
Phylum: Chordata
Class: Mammalia
Order: Diprotodontia
Family: Potoroidae
Genus: Potorous
Behavior: Herbivore/
Nocturnal

Northern Hairy-nosed Wombat

Lasiorhinus krefftii

THIS MARSUPIAL'S SKULL is very similar to that of the placental rodents elsewhere in the world. Wombats have a pair of large, robust incisors in both the upper and lower jaws (like a beaver) anchored deep in the jawbone. They have no canines, and a wide diastema. The hairy-nosed wombat is one of three wombat species and slightly larger than the common wombat. As a marsupial, it has a pouch—in this case one that opens backward so that dirt doesn't get in when it digs.

AKA: Yaminon
Kingdom: Animalia
Phylum: Chordata
Class: Mammalia
Order: Diprotodontia
Family: Vombatidae
Genus: Lasiorhinus
Behavior: Herbivore/
Nocturnal

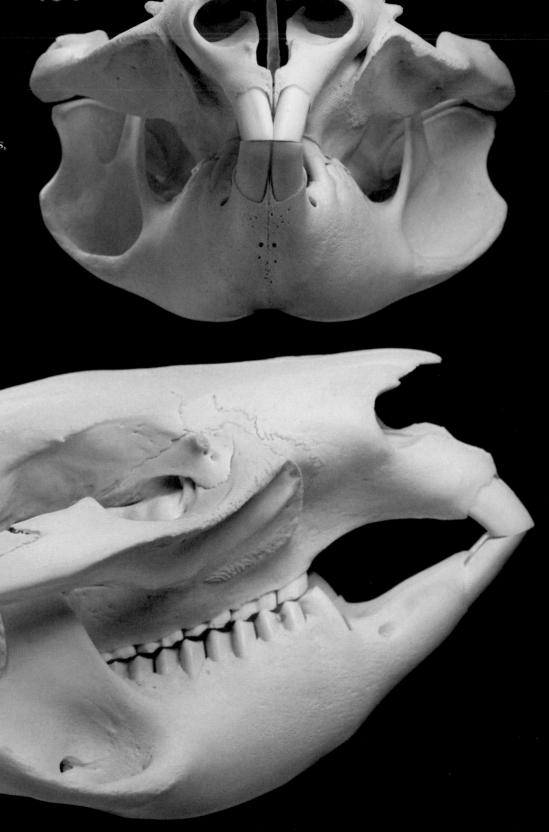

European Mole
Talpa europaea ▷

THE EUROPEAN MOLE SKULL, here seen mounted open, is a fearsome thing—if you are a worm or small insect that finds its way into its path. Since European moles are almost exclusively subterranean, there are no discernible orbits. The associated skeleton, one of just a few in Alan Dudley's collection, is in fact that of an eastern mole. It has huge front feet, like miniature snow shovels (useful for burrowing below your garden or the golf course!).

AKA: Common Mole
Kingdom: Animalia
Phylum: Chordata
Class: Mammalia
Order: Soricomorpha

Family: Talpidae
Genus: Talpa
Behavior: Carnivore/
Subterranean

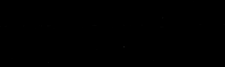

Common Shrew
Sorex araneus ◁

IF ANY SPECIES COULD be considered the blueprint for the early mammal in its original and basic form, it is the shrew. Shrews are not rodents; they have sharp, spikelike teeth, unlike any rodent that they may superficially be confused with. They are found all over the world, and some species are venomous. Shrews have characteristic red-tipped teeth and a very high metabolic rate.

AKA: Eurasian Shrew
Kingdom: Animalia
Phylum: Chordata
Class: Mammalia

Order: Soricomorpha
Family: Soricidae
Genus: Sorex
Behavior: Insectivore/Nocturnal

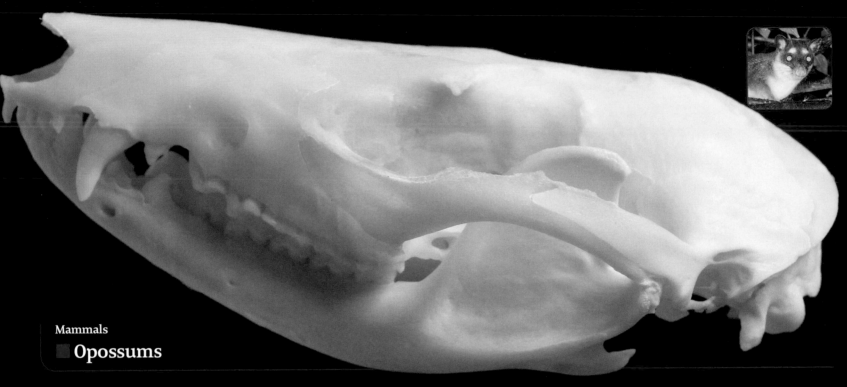

Brown Four-eyed Opossum

Metachirus nudicaudatus △

THIS IS AN ELEGANT SKULL, with sharp teeth and a sagittal crest. The brown four-eyed opossum is a marsupial found in South America. Of course, it doesn't have four eyes, but it does sport two white spots on its forehead.

Kingdom: Animalia
Phylum: Chordata
Class: Mammalia
Order: Didelphimorphia

Family: Didelphidae
Genus: Metachirus
Behavior: Omnivore/ Nocturnal

Virginia Opossum

Didelphis virginiana ◁

THE VIRGINIA OPOSSUM is the only marsupial (it raises its young in a pouch, kangaroo-style) that occurs naturally in North America. Its skull is thus prized among skull collectors. Famously, it will "play possum" or feign death when threatened.

AKA: North American Opossum
Kingdom: Animalia
Phylum: Chordata
Class: Mammalia
Order: Didelphimorphia
Family: Didelphidae
Genus: Didelphis
Behavior: Omnivore/Diurnal

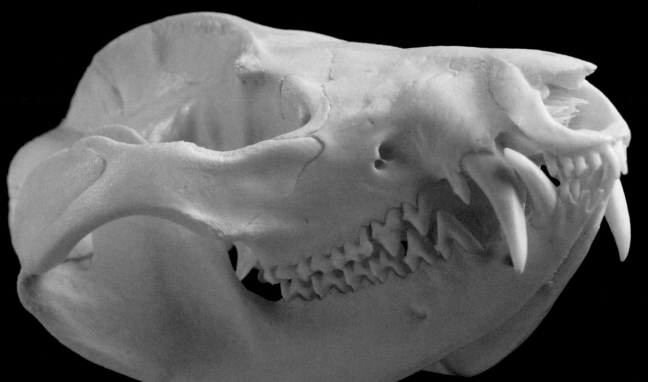

Greater Galago
Otolemur sp. ▷

THIS SKULL IS NOT very different from the other galago skull in the collection, although it does have a longer rostrum and relatively smaller eyes. At night, this animal forages for small insects and slugs but also eats berries and seeds—as well as small birds and reptiles, when opportunity arises. This is one of those species that keeps getting divided into subspecies (not uncommon among small nocturnal primates in Africa), so it may soon be in for another name change. The photograph is of a brown greater galago.

AKA: Thick-tailed Bushbaby
Kingdom: Animalia
Phylum: Chordata
Class: Mammalia

Order: Primates
Family: Galagonidae
Genus: Otolemur
Behavior: Omnivore/ Nocturnal

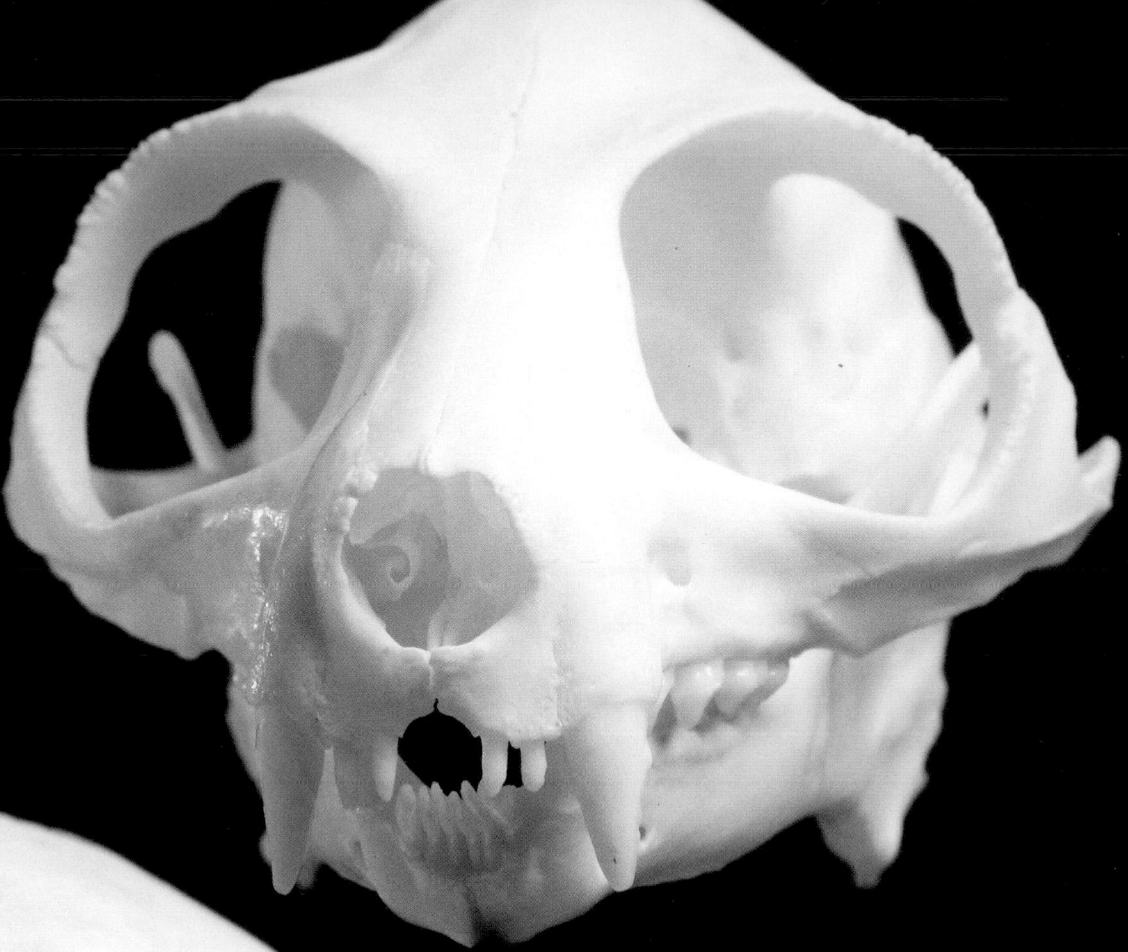

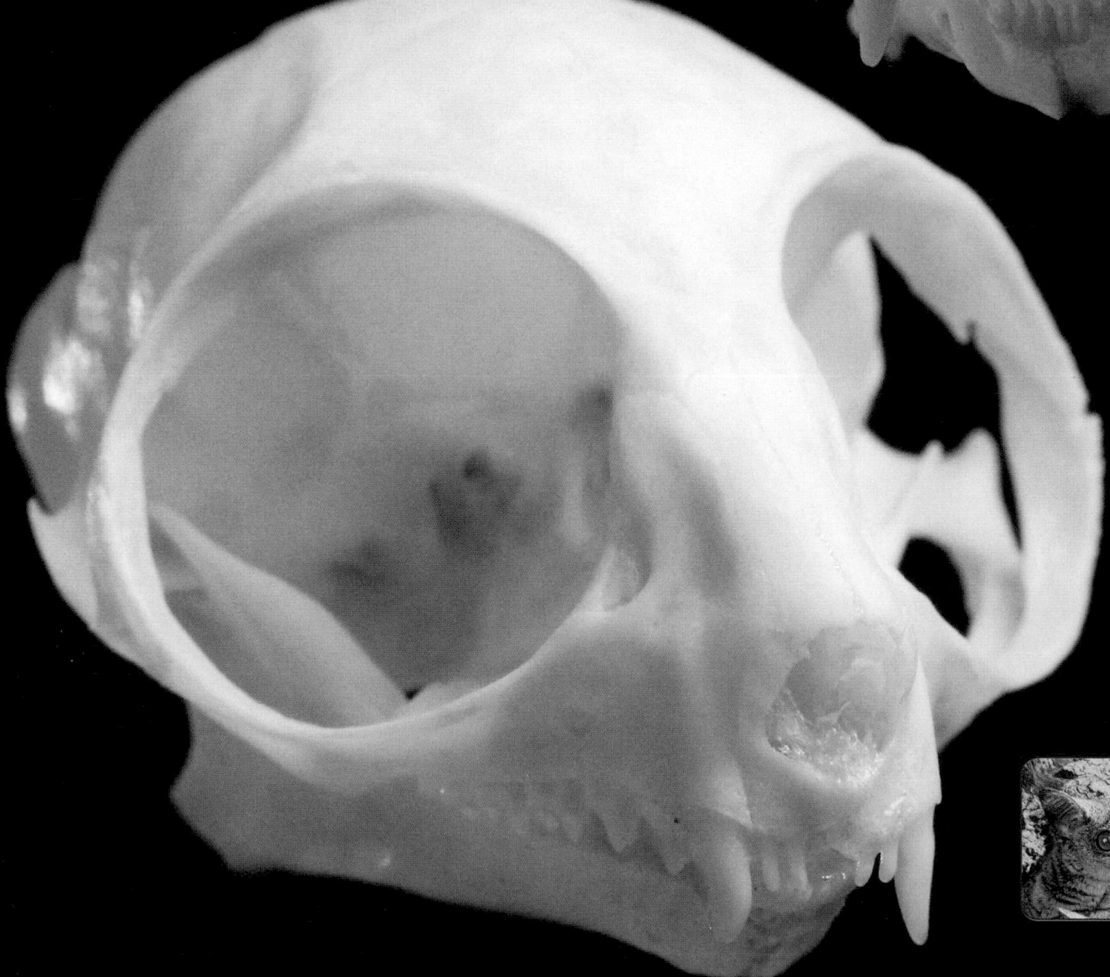

Senegal Galago
Galago senegalensis ◁

GALAGOS ARE SOMETIMES known as bushbabies due to their haunting night cries. These nocturnal primates have huge orbits to house eyes large enough to see in the dark. Senegal galagos have a mixed diet of insects and acacia gum.

AKA: Lesser Bushbaby
Kingdom: Animalia
Phylum: Chordata
Class: Mammalia
Order: Primates
Family: Galagonidae
Genus: Galago
Behavior: Omnivore/ Nocturnal

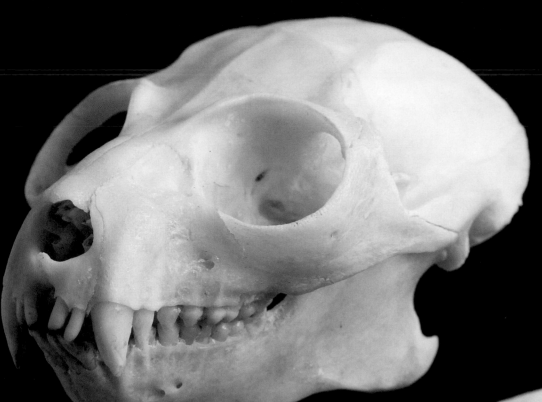

Potto
Perodicticus potto ◁

THE POTTO SKULL IS UNUSUAL—its bulging orbits are enormous, appearing almost goggle-like. A nocturnal animal with a short rostrum, the potto lives among the trees in the equatorial jungles of central Africa. It has a strong jaw, which allows it to feed on tree gum as well as fruits and insects.

AKA: Softly-softly
Kingdom: Animalia
Phylum: Chordata
Class: Mammalia
Order: Primates

Family: Loridae
Genus: Perodicticus
Behavior: Frugivore/
Nocturnal

Philippine Tarsier
Tarsius syrichta Maumag ▷

THE PHILIPPINE TARSIER has undoubtedly the most impressive eye orbits of all the skulls in the collection. Two-thirds of the skull volume can be attributed to the eyes of this small nocturnal primate. Each eye is larger than its brain, and the neurologic wiring of the eyes to the brain distinguishes it taxonomically from the rest of the primates.

Kingdom: Animalia
Phylum: Chordata
Class: Mammalia
Order: Primates
Family: Tarsiidae
Genus: Tarsius
Behavior: Insectivore/
Nocturnal

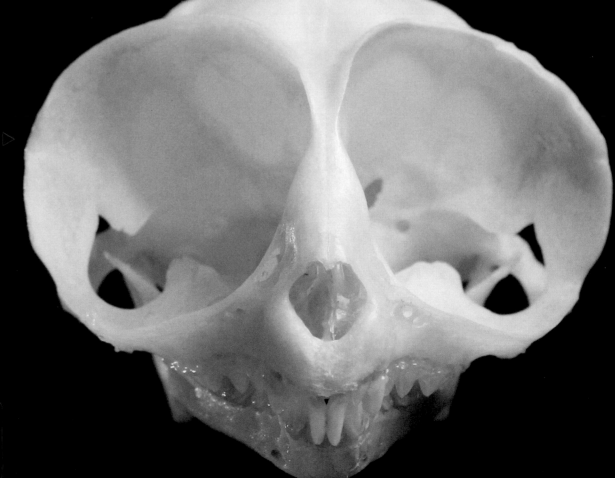

Siamang
Hylobates syndactylus ◁

THE SIAMANG is a gibbon: a group of apes found only in Southeast Asia. Its skull is typically primate, with forward-facing orbits accommodating binocular vision. Despite the fierce-looking teeth, the siamang is almost entirely vegetarian. Siamangs communicate by song, amplifying the sounds with a throat sac.

Kingdom: Animalia
Phylum: Chordata
Class: Mammalia
Order: Primates
Family: Hylobatidae
Genus: Hylobates
Behavior: Frugivore/
Diurnal

Bonobo
Pan paniscus ▷

MANY PEOPLE WOULD be surprised to learn that there are two species of chimpanzee—living on opposite sides of the Congo River. Once known as the pygmy chimpanzee, the bonobo is by far the rarer and less familiar of the two. This bonobo skull is probably the closest match to the human skull in the animal world. The bonobo's head is smaller than that of the common chimpanzee, with less prominent brow ridges.

AKA: Pygmy
Chimpanzee
Kingdom: Animalia
Phylum: Chordata
Class: Mammalia

Order: Primates
Family: Hominidae
Genus: Pan
Behavior: Frugivore/
Diurnal

DUDLEY'S NOTES:
The bonobo is a smaller version of a chimpanzee and the two skulls can be difficult to tell apart. This is a very old specimen and it's had a lot of repairs made to it.

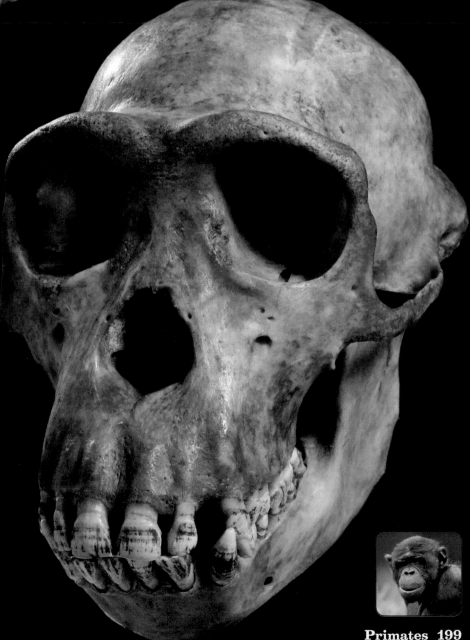

Lowland Gorilla
Gorilla gorilla

GORILLAS HAVE THE largest heads of any living primate, characterized by prominent brow ridges and, in the males, a large sagittal crest. This is the skull of a female, probably very old (judging from its ivory color); there is just a hint of the crest that is so distinctive among the males. Like humans and the other apes, adult lowland gorillas have thirty-two teeth, which are robust and mounted in powerful jaws, allowing this herbivore to grind up its food. The full skeleton on page 121 is that of a male.

AKA: Western Gorilla
Kingdom: Animalia
Phylum: Chordata
Class: Mammalia
Order: Primates

Family: Hominidae
Genus: Gorilla
Behavior: Herbivore/
Diurnal

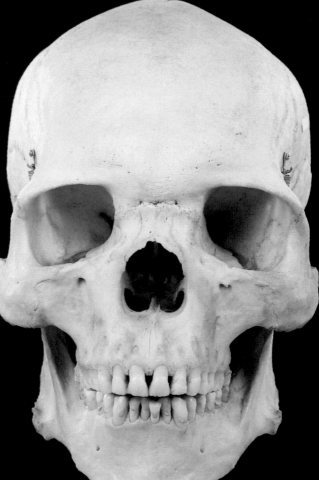

Human
Homo sapiens ▷

OUR SPECIES HAS been around only for about 200,000 years and for most of that time exclusively in Africa. Then, some 70,000 years ago, our ancestors began to explore the world at large, over the next 30,000 years—except for the Americas, which were colonized just 15,000 to 20,000 years ago. The photographic example of *Homo sapiens* is Simon Winchester, the author.

Kingdom: Animalia
Phylum: Chordata
Class: Mammalia
Order: Primates

Family: Hominidae
Genus: Homo
Behavior: Omnivore/
Diurnal

DUDLEY'S NOTES:
Lowland gorillas are very difficult to get a hold of these days. Most of them go to museums. This gorilla is an antique and it's a female. The males are much more impressive, but when a gorilla skull comes your way you really have to take it because you never know when the chance will come again.

Orangutan
Pongo pygmaeus

THESE SKULLS demonstrate the extent of sexual dimorphism among the great apes: males have a more prominent sagittal crest, wider zygomatic arches, and larger canines. Orangutans are the only great ape found in Southeast Asia and have arms twice as long as their legs.

Kingdom: Animalia
Phylum: Chordata
Class: Mammalia
Order: Primates
Family: Hominidae
Genus: Pongo
Behavior:
Omnivore/
Diurnal

Red Ruffed Lemur
Varecia variegata ∧

THE RED RUFFED LEMUR is the largest typical lemur and is resident on the east coast of Madagascar. This skull is dominated by the large orbits and prominent canine teeth. Although not visible here, the lower incisors tilt forward at about forty-five degrees and are used as a specialized "tooth comb" during mutual grooming.

Kingdom: Animalia
Phylum: Chordata
Class: Mammalia
Order: Primates
Family: Lemuridae
Genus: Varecia
Behavior: Omnivore/ Diurnal

Ring-tailed Lemur
Lemur catta ∧

THE SKULL OF THIS Madagascan icon resembles that of the potto, to which it is closely related. It has flared and prominent orbits (reflecting its good night vision) and prominent canine teeth at the end of a fox-like muzzle. There are twenty-two species of lemur, all found in Madagascar.

Kingdom: Animalia
Phylum: Chordata
Class: Mammalia
Order: Primates
Family: Lemuridae
Genus: Lemur
Behavior: Omnivore/ Diurnal

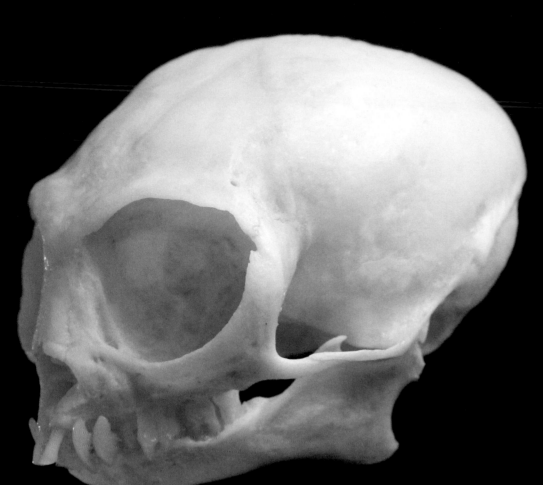

Pygmy Marmoset
Callithrix pygmaea ◁

ONE OF THE SMALLEST primates, the pygmy marmoset is native to South American rain forests in Peru, Brazil, and Colombia. Its tiny skull is dominated by the eyes and braincase. Its buck teeth (that lend the skull a comical appearance) are an adaptation for gouging tree bark to access the gum on which it feeds.

AKA: Dwarf Monkey
Kingdom: Animalia
Phylum: Chordata
Class: Mammalia
Order: Primates

Family: Callitrichidae
Genus: Callithrix
Behavior: Omnivore/ Diurnal

Red-handed Tamarin
Saguinus midas ▷

THIS SKULL HAS the same comical buck teeth seen in the related pygmy marmoset, but differs in having a blunt face and wide-flaring but delicate zygomatic arches. It feeds on gum, and the incisors are adapted for scoring the bark on trees. When living, it has, as its scientific name suggests, golden feet.

AKA: Golden-handed Tamarin
Kingdom: Animalia
Phylum: Chordata
Class: Mammalia
Order: Primates
Family: Callitrichidae
Genus: Saguinus
Behavior: Omnivore/ Diurnal

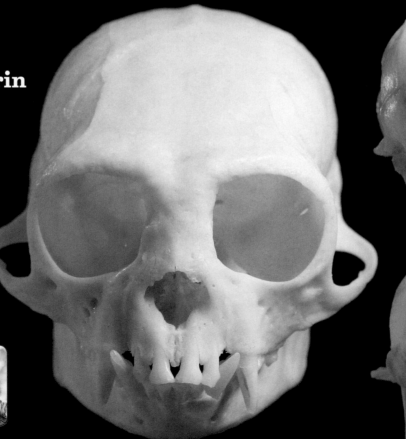

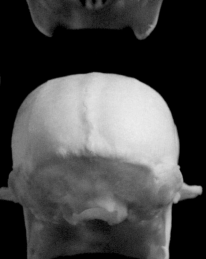

Hamadryas Baboon
Papio hamadryas

THE HAMADRYAS BABOON has a long doglike muzzle; its dentition and forward-facing eyes are indications that it is a primate. The hamadryas was sacred to the ancient Egyptians; skulls such as this have been found in their tombs.

AKA: Sacred Baboon
Kingdom: Animalia
Phylum: Chordata
Class: Mammalia
Order: Primates
Family: Cercopithecidae
Genus: Papio
Behavior: Herbivore/Diurnal

Mandrill
Mandrillus sphinx

OFTEN CONFUSED with the closely related baboon, the mandrill is the world's largest monkey. Famous for its colorful muzzle, the underlying structures can be clearly seen in the long rostrum of this skull. It brandishes a set of canines fit for any predator. Mandrills live in large troops of up to 100 individuals that are dominated by the males with the biggest canines. So think of this three-inch-long sexually selected weaponry as a kind of monkey antler. Female skulls are slighter, with much smaller canines.

Kingdom: Animalia
Phylum: Chordata
Class: Mammalia
Order: Primates
Family: Cercopithecidae
Genus: Mandrillus
Behavior: Omnivore/Diurnal

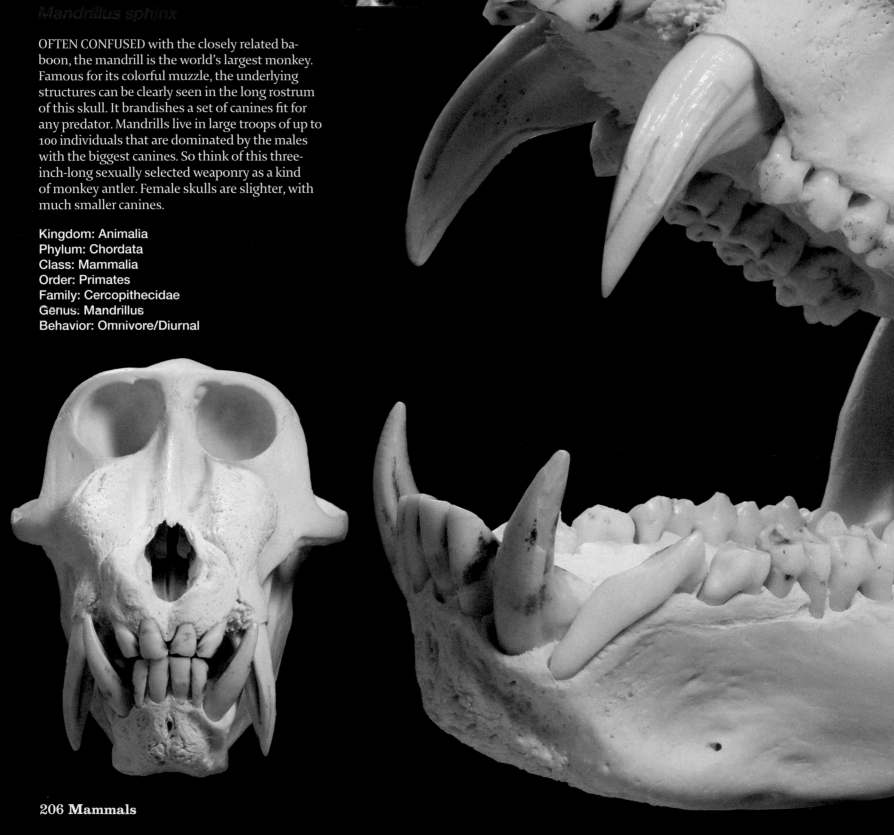

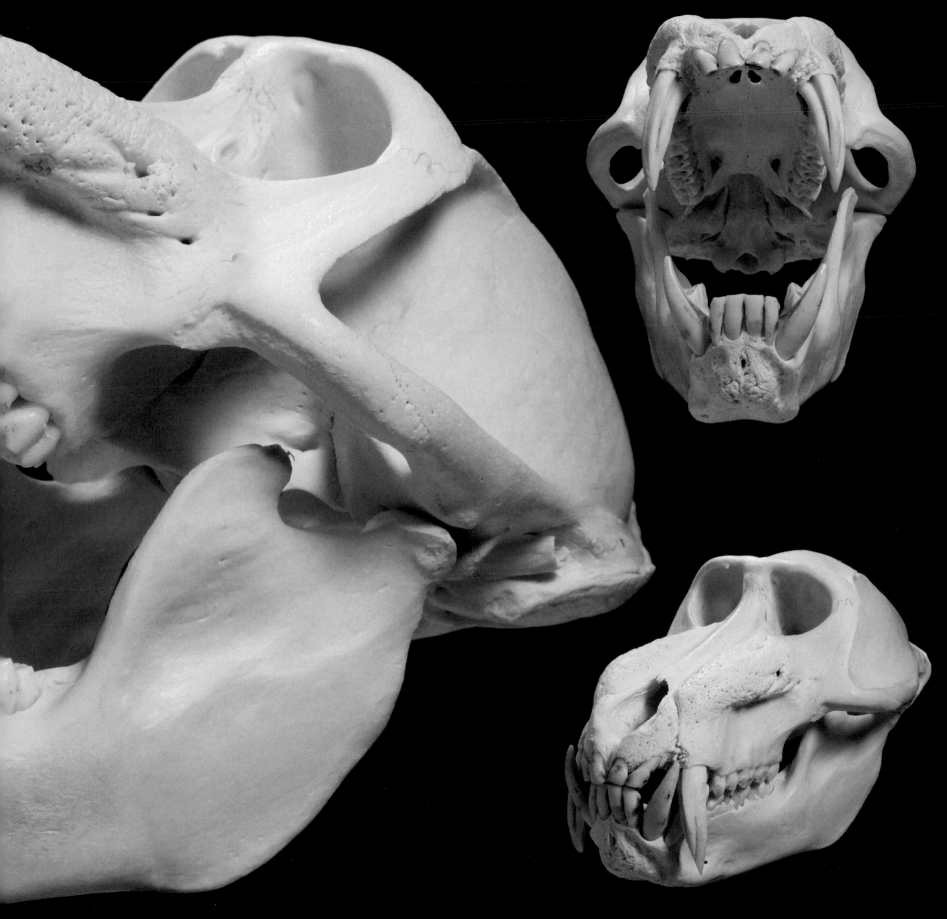

Black-and-White Colobus
Colobus sp. ▷

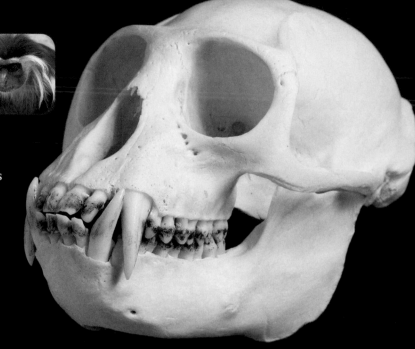

THIS MONKEY HAS a typical primate skull: large forward-facing eyes and a large cranium. Like many other primates, the male colobus has an impressive set of canines. These are used in status displays and also as defense—many colobus are actively hunted by chimpanzees. There are five species of black-and-white colobus monkey.

Kingdom: Animalia
Phylum: Chordata
Class: Mammalia
Order: Primates
Behavior: Frugivore/Diurnal

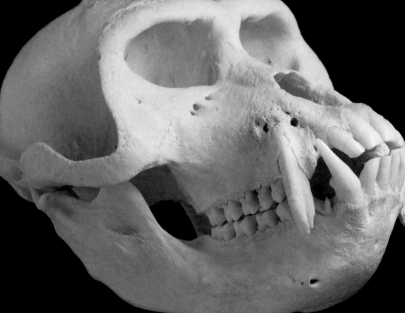

Stump-tailed Macaque
Macaca arctoides ◁

IN THE FLESH, these are not the prettiest of primates. Aptly named from their short, stumpy tails, these macaques are also unusual in that, like some of us, they go bald in old age. They spend a great deal of time on the ground in the forests of southern China and its border countries. The skull is characterized by wide-flaring zygomatic arches and a narrow mandible. The large canine teeth are used during dominance displays by males.

AKA: Bear Macaque
Kingdom: Animalia
Phylum: Chordata
Class: Mammalia
Order: Primates
Family: Cercopithecidae
Genus: Macaca
Behavior: Omnivore/Diurnal

Tufted Gray Langur
Semnopithecus priam ▷

UNUSUALLY FOR MONKEYS, this species from southern India is primarily a leaf eater. This skull lacks the prominent canine teeth of other monkeys. Forward-facing eyes, making binocular vision possible, are a distinguishing feature of primates, giving them a recognizable face.

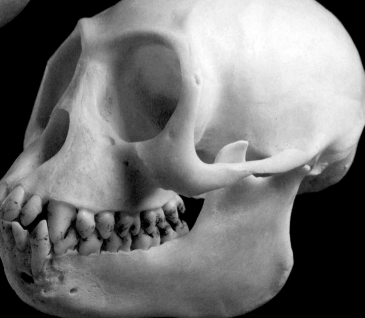

AKA: Crested Gray Langur
Kingdom: Animalia
Phylum: Chordata
Class: Mammalia
Order: Primates
Family: Cercopithecidae
Genus: Semnopithecus
Behavior: Herbivore/Diurnal

Vervet Monkey

Chlorocebus pygerythrus ▷

THE VERVET MONKEY is widespread in southern and eastern Africa. Males dominate the troop; competition among them can at times be fierce, with those lethal canines inflicting serious and occasionally life-threatening wounds on each other. They are highly adaptable animals and are as much at home in town suburbs as they are on the open savanna.

Kingdom: Animalia
Phylum: Chordata
Class: Mammalia
Order: Primates
Family: Cercopithecidae
Genus: Chlorocebus
Behavior: Omnivore/ Diurnal

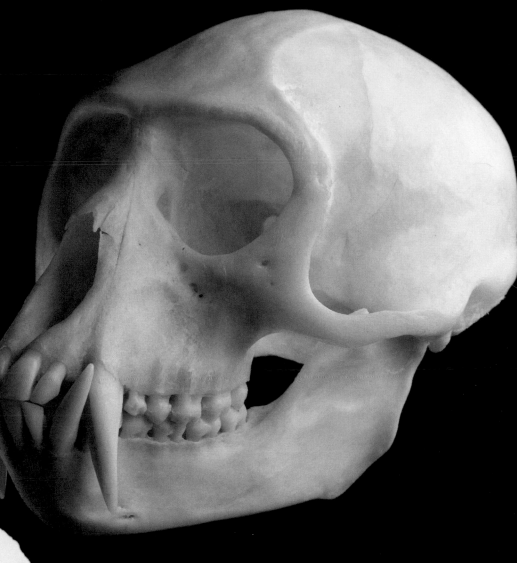

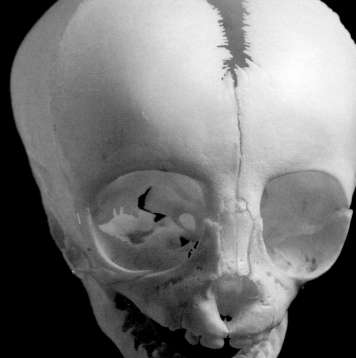

Black-headed Spider Monkey

Ateles fusciceps ◁

THIS SKULL IS PERHAPS one of the more disturbing in the collection as it bears a close resemblance to the skull of a small child. You can see that the sutures of the skull have yet to close, and the fontanelle is clearly visible. Unlike an immature human skull the teeth have already emerged.

Kingdom: Animalia
Phylum: Chordata
Class: Mammalia
Order: Primates

Family: Cebidae
Genus: Ateles
Behavior: Frugivore/ Diurnal

European Hare
Lepus europaeus

LIKE THOSE OF RABBITS, European hare skulls are typified by paired incisors separated from their molars by a gap known as a diastema. Neither are rodents; they were reclassified as lagomorphs in 1912. They are distinctive in having skulls that are fenestrated; that is, they possess openings that are thought to allow cooling of the blood. The rostrum is large and wide, and the bone above the eyes has small characteristic processes that point to the front and rear. The auditory bullae are large, accommodating their acute hearing.

AKA: Brown Hare
Kingdom: Animalia
Phylum: Chordata
Class: Mammalia
Order: Lagomorpha

Family: Leporidae
Genus: Lepus
Behavior: Herbivore/
Crepuscular

European Rabbit
Oryctolagus cuniculus

In this unusual specimen, the incisors (which grow throughout the rabbit's life, as do those of rodents) have failed to align and grind against each other. Had the animal had sufficiently hard material to gnaw on during its lifetime, the teeth would have worn naturally. Rather like the eccentric long nails sported by Indian mystics, this animal's teeth twisted and curved as they grew. It is hard to see how this particular rabbit could have eaten properly—or how its owner could have failed to notice its abnormality.

AKA: Common Rabbit
Kingdom: Animalia
Phylum: Chordata
Class: Mammalia
Order: Lagomorpha
Family: Leporidae
Genus: Oryctolagus
Behavior: Herbivore/Crepuscular

DUDLEY'S NOTES:
I have several rabbits in my collection, but this is a particular favorite. It has abnormally long teeth because when the rabbit was alive the teeth never actually connected when it chewed its food. Consequently, the teeth never ground down and instead kept growing. I love the abnormality.

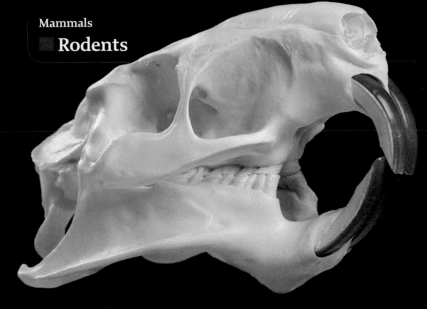

Coypu
Myocastor coypus ◁

AT A GLANCE, the coypu (a native of South America, also known as the nutria) looks much like a very large water rat. The skull can be identified by its bright, dark orange incisor teeth (the beaver also has pigmented teeth, but they are a lighter shade). There was a huge, dedicated, and ultimately successful eradication program in Britain's East Anglia in the late 1980s after animals escaped from farms and bred in the wild.

AKA: Nutria
Kingdom: Animalia
Phylum: Chordata
Class: Mammalia

Order: Rodentia
Family: Echimyidae
Genus: Myocastor
Behavior: Herbivore/Nocturnal

Long-tailed Chinchilla
Chinchilla lanigera ▷

THE LONG-TAILED CHINCHILLA, a native of South America, was once prized for its soft, dense fur but is now more popular as a household pet. It has the characteristic skull of a rodent and large auditory bullae. The incisors on this skull are overgrown, curling round into the palate.

AKA: Chilean Chinchilla
Kingdom: Animalia
Phylum: Chordata
Class: Mammalia
Order: Rodentia
Family: Chinchillidae
Genus: Chinchilla
Behavior: Omnivore/
Nocturnal

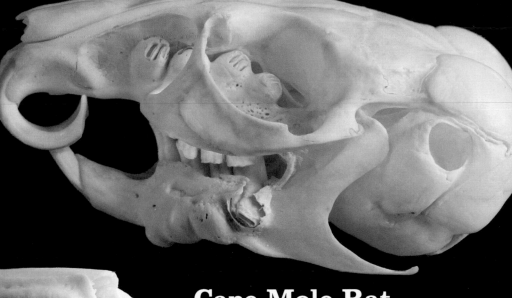

Cape Mole Rat
Georychus capensis ◁

THIS MOLE-LIKE RODENT has grooved gnawing teeth used not only for nibbling vegetation, but also for digging. The Cape mole rat's lips close behind its incisors to stop soil from entering its mouth. Its head is large relative to its body, but its eyes are small, as it lives a subterranean life. Unlike its more famous and furless relation, the naked mole rat, the Cape mole rat is a largely solitary animal. Although much of its habitat has been destroyed, it is now finding a new home under golf courses.

Kingdom: Animalia
Phylum: Chordata
Class: Mammalia
Order: Rodentia

Family: Bathyergidae
Genus: Georychus
Behavior: Herbivore/
Subterranean

Mountain Beaver

Aplodontia rufa ◁

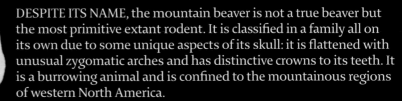

DESPITE ITS NAME, the mountain beaver is not a true beaver but the most primitive extant rodent. It is classified in a family all on its own due to some unique aspects of its skull: it is flattened with unusual zygomatic arches and has distinctive crowns to its teeth. It is a burrowing animal and is confined to the mountainous regions of western North America.

AKA: Sewellel Beaver
Kingdom: Animalia
Phylum: Chordata
Class: Mammalia
Order: Rodentia
Family: Aplodontiidae
Genus: Aplodontia
Behavior: Herbivore/
Nocturnal

North American Beaver

Castor canadensis ▷

THIS NORTH AMERICAN BEAVER SKULL, which has twenty teeth, is large and heavy with a stout rostrum, heavy zygomatic arches, and a long and narrow braincase with well-developed V-shaped temporal ridges. It uses its long, strong incisors to girdle and fell large deciduous trees to build dams. The beaver has eyes designed to see underwater and a warm double coat of fur that is rendered waterproof with a substance exuded near its tail.

Kingdom: Animalia
Phylum: Chordata
Class: Mammalia
Order: Rodentia
Family Castoridae

Genus: Castor
Behavior: Herbivore/
Nocturnal

North African Crested Porcupine

Hystrix cristata ◁

THIS PORCUPINE IS FOUND throughout Africa and in parts of southern Europe. Its skull is quite specialized: its infraorbital foramen (for chewing muscle attachment) and nasal cavity (since the animal, often nocturnal, hunts by smell) are greatly enlarged. It also has a noteworthy habit of bone collecting (possibly to provide a source of potassium), and its ancient burrows, when unearthed, are often a treasure trove for archaeologists and paleontologists—and skull collectors!

Kingdom: Animalia
Phylum: Chordata
Class: Mammalia
Order: Rodentia

Family: Hystricidae
Genus: Hystrix
Behavior: Herbivore/
Nocturnal

Lowland Paca

Cuniculus paca ▷

THIS LOWLAND PACA is a rain forest rodent with swollen cheeks that are reflected in the skull's structure. You can see the high degree of pneumatization of the zygomatic region, which, in layman's terms is the coral-like surface of the cheek bones. Behaviorists argue that it serves to amplify the teeth-grinding sounds that males of the species make. It is a versatile creature of the forests of South and Central America able to swim well and climb trees in search of fruit.

AKA: Spotted Paca
Kingdom: Animalia
Phylum: Chordata
Class: Mammalia
Order: Rodentia
Behavior: Frugivore/
Nocturnal

DUDLEY'S NOTES:
The paca is my favorite rodent. What I like about them is they've got massive cheekbones that come right around the face and look a lot like Roman headgear.

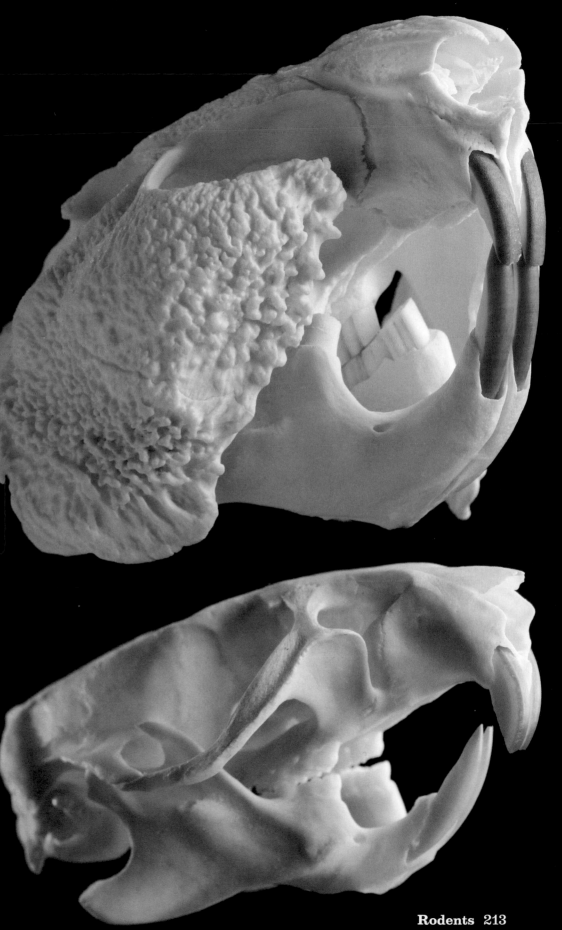

Brown Rat

Rattus norvegicus ▷

LIKE LARGER HERBIVORES, rats and other rodents have a diastema—a space that separates the incisors from the molars along the jaw when seen in profile. If you find this skull in your attic or basement, call pest control!

AKA: Common Rat,
Sewer Rat
Kingdom: Animalia
Phylum: Chordata
Class: Mammalia
Order: Rodentia
Family: Muridae
Genus: Rattus
Behavior: Omnivore/
Nocturnal

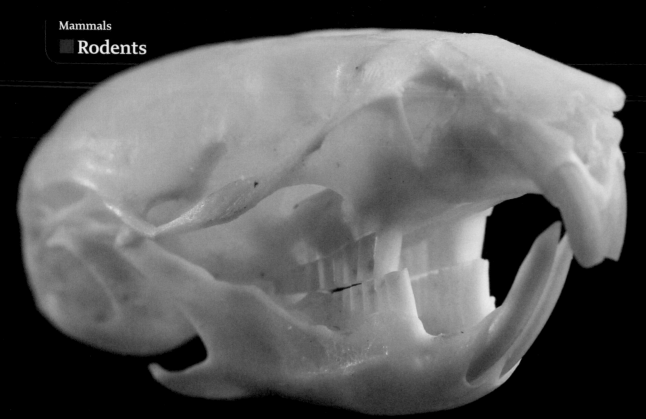

Field Vole
Microtus agrestis ◁

THE FIELD VOLE'S SKULL is tiny and delicate. One can easily differentiate the skull of this vole from that of the common shrew by the large diastema, and the prominent cheek bones.

Kingdom: Animalia
Phylum: Chordata
Class: Mammalia
Order: Rodentia
Family: Cricetidae
Genus: Microtus
Behavior: Omnivore/ Nocturnal

Capybara
Hydrochaeris hydrochaeris ▷

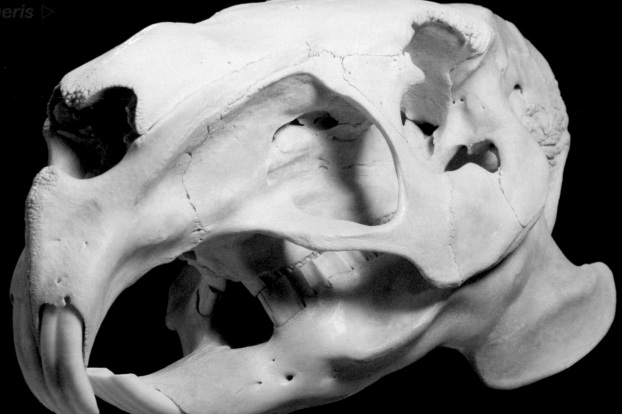

THE CAPY- BARA is the largest living rodent, nearly the size of a small antelope. It is found throughout large areas of lowland South America, wherever there is wetland. There are now small escaped populations in several European countries. Its jaw articulation is nonperpendicular: this means that instead of grinding its food from side to side, as do other herbivores, it grinds its food back and forth. *capybara* means "master of the grasses."

Kingdom: Animalia
Phylum: Chordata
Class: Mammalia
Order: Rodentia
Family: Hydrochaeridae
Genus: Hydrochaeris
Behavior: Herbivore/Diurnal

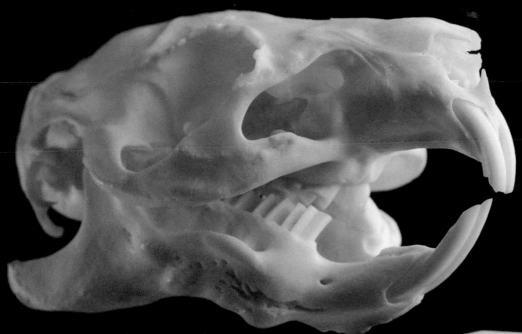

Guinea Pig
Cavia porcellus ◁

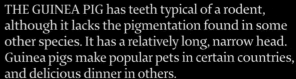

THE GUINEA PIG has teeth typical of a rodent, although it lacks the pigmentation found in some other species. It has a relatively long, narrow head. Guinea pigs make popular pets in certain countries, and delicious dinner in others.

AKA: Cavy
Kingdom: Animalia
Phylum: Chordata
Class: Mammalia
Order: Rodentia

Family: Caviidae
Genus: Cavia
Behavior: Herbivore/
Diurnal

Patagonian Mara
Dolichotis patagonum ▷

THE PATAGONIAN MARA is closely related to the guinea pig. Its skull is that of a typical rodent with a huge diastema; the molars are tucked away deep in the mouth. Maras are social animals and reputed to make good pets. They are found in open and semiarid areas of Argentina.

AKA: Patagonian Hare
Kingdom: Animalia
Phylum: Chordata
Class: Mammalia
Order: Rodentia
Family: Caviidae
Genus: Dolichotis
Behavior: Herbivore/
Diurnal

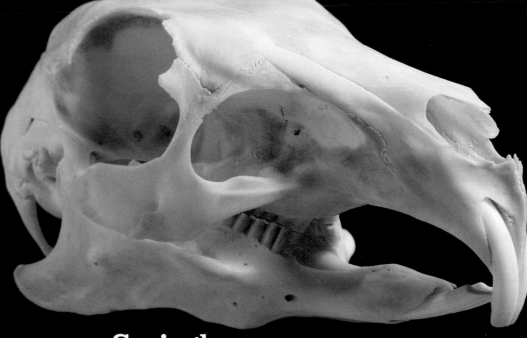

Springhare
Pedetes capensis ◁

DESPITE ITS NAME, the springhare is a rodent and not related to hares at all. It hops around southeastern Africa, much like a small kangaroo; hence the "spring" in its name. The skull is curious, and you might not be sure which way round it should face until you pick out the rodent teeth below. The springhare is a crepuscular animal with excellent hearing, evidenced by its well-developed auditory bullae.

AKA: Springhaas
Kingdom: Animalia
Phylum: Chordata
Class: Mammalia
Order: Rodentia

Family: Pedetidae
Genus: Pedetes
Behavior: Herbivore/
Crepuscular

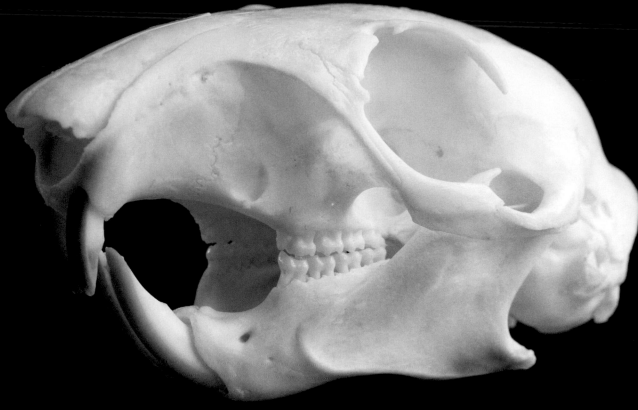

Gray Squirrel
Sciurus carolinensis ◁

THE GRAY SQUIRREL SKULL is typical of rodents and has a bright orange glaze to the front of its teeth—like a small version of the coypu. The gray squirrel is native to eastern America, but a burgeoning feral population exists in Britain.

Kingdom: Animalia
Phylum: Chordata
Class: Mammalia
Order: Rodentia
Family: Sciuridae
Genus: Sciurus
Behavior: Herbivore/ Diurnal

Woodchuck
Marmota monax ▷

THIS SQUIRREL-LIKE RODENT is a member of the marmot family and has a broad, flattened skull typical of burrowing animals. They are prodigious tunnelers and the scourge of farmers.

AKA: Groundhog
Kingdom: Animalia
Phylum: Chordata
Class: Mammalia
Order: Rodentia
Family: Sciuridae
Genus: Marmota
Behavior:
Herbivore/
Diurnal

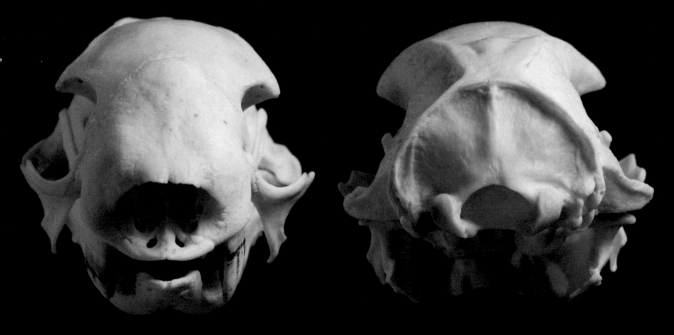

Linnaeus's Two-toed Sloth
Choloepus didactylus

THIS IS A CURIOUS ANIMAL with a curious skull. The Linnaeus's two-toed sloth lacks true tympanic bullae. There are, instead, what anatomists call ectotympanic bullae below an incomplete zygomatic arch. No incisors are present, but there are flat-topped molars for grinding plant material. South America is home to several species of sloth, both two- and three-toed.

AKA: Southern Two-toed Sloth, Unau
Kingdom: Animalia
Phylum: Chordata
Class: Mammalia
Order: Xenarthra
Family: Megalonychidae
Genus: Choloepus
Behavior: Herbivore/Nocturnal

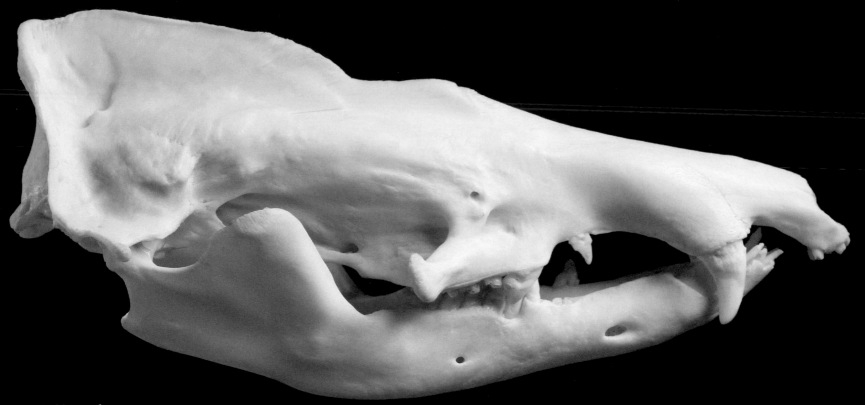

■ Tenrecs

Common Tenrec

Tenrec ecaudatus △

TENRECIDAE ARE A highly diverse and populous family of animals found on Madagascar. This skull is from a very old specimen obtained from a museum. It is hard to pin it down to a specific species (although its Dutch label does describe it as a "gewone tenrec" or "common tenrec"). Despite its small size, it has a pronounced sagittal crest and flaring at the rear of the skull. The common tenrec is among the largest of the family; despite a relative paucity of teeth, it feeds on frogs and small mammals.

AKA: Tailless Tenrec
Kingdom: Animalia
Phylum: Chordata
Class: Mammalia
Order: Afrosoricida

Family: Tenrecidae
Genus: Tenrec
Behavior: Insectivore/
Nocturnal

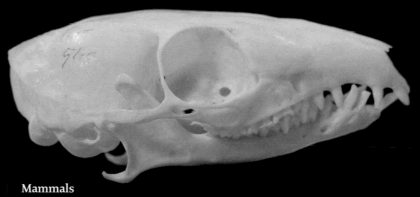

Mammals

■ Treeshrews

Common Treeshrew

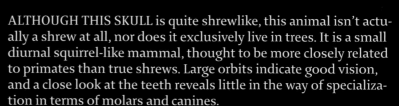

Tupaia glis △

ALTHOUGH THIS SKULL is quite shrewlike, this animal isn't actually a shrew at all, nor does it exclusively live in trees. It is a small diurnal squirrel-like mammal, thought to be more closely related to primates than true shrews. Large orbits indicate good vision, and a close look at the teeth reveals little in the way of specialization in terms of molars and canines.

Kingdom: Animalia
Phylum: Chordata
Class: Mammalia
Order: Scandentia

Family: Tupaiidae
Genus:Tupaia
Behavior: Carnivore/
Diurnal

Rock Hyrax
Procavia capensis

THE ROCK HYRAX SKULL is often featured in anatomy collections because of its close taxonomic relationship to elephants and manatees. It has long, tusklike incisors and a set of molars similar to that of the rhino. Its mandible is large, more than half the skull by bone volume.

AKA: Cape Hyrax
Kingdom: Animalia
Phylum: Chordata
Class: Mammalia
Order: Hyracoidea
Family: Procaviidae
Genus: Procavia
Behavior: Herbivore/ Diurnal

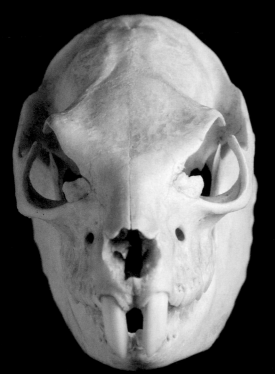

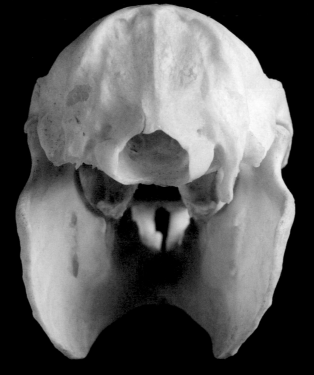

DUDLEY'S NOTES:
The rock hyrax is such a strange-looking animal. I've always wondered what the skull looked like. When I finally acquired this one it was from an antique dealer and it had one tooth missing. The two front teeth are quite long and have a big gap in between them.

African Elephant

Loxodonta africana

THIS SKULL IS HELD by the Field Museum in Chicago, and a tag attached to it suggests an interesting provenance: the Ringling Brothers Circus. Elephant skulls are steeped in myth and folklore. It is said that the legend of the Cyclops arose from the discovery of dwarf elephant skulls on Mediterranean islands. Looking at the skull, it is easy to see how such legends could have arisen: the grinning mouth with a protruding chin and the single hole in the forehead (from which the trunk protrudes) in a skull that was twice the size of a human's could suggest a terrifying creature.

Kingdom: Animalia
Phylum: Chordata
Class: Mammalia
Order: Proboscidea
Family: Elephantidae
Genus: Loxodonta
Behavior: Herbivore/
Diurnal and Nocturnal

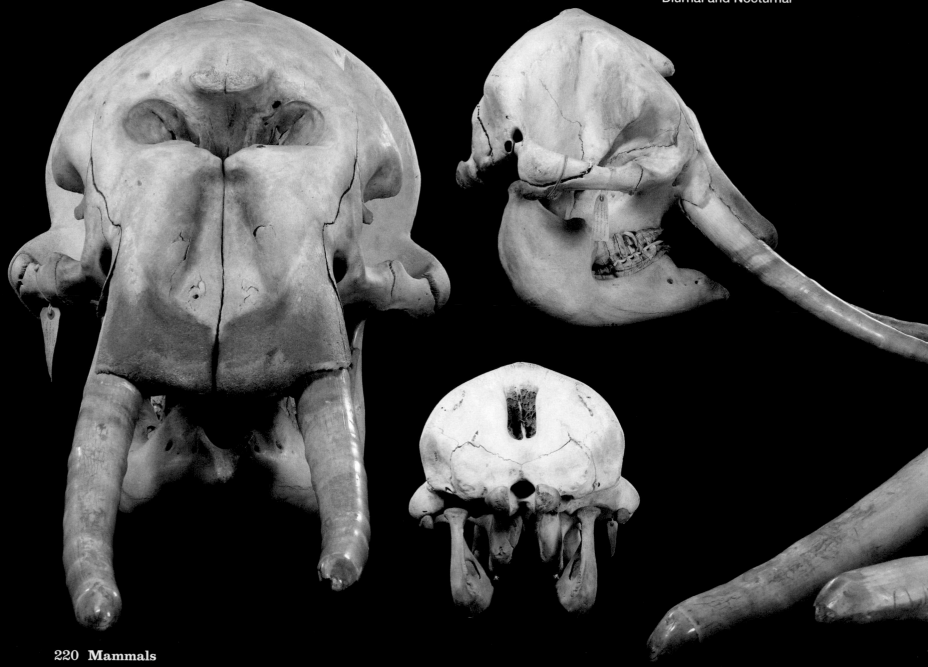

Common Bottlenosed Dolphin
▷ *Tursiops truncatus*

DESPITE BEING A MAMMAL, the dolphin's rostrum is commonly referred to as a beak (so we hear of beaked whales, short-beaked dolphins, and so on). Unlike most terrestrial mammals, dolphins have practically no differentiation among their rows of conical teeth, which thus look distinctly reptilian. Dolphins are among the most intelligent of mammals, evidenced, in part, by a braincase substantially larger than our own.

AKA: Franciscana
Kingdom: Animalia
Phylum: Chordata
Class: Mammalia
Order: Cetacea

Family: Delphinidae
Genus: Tursiops
Behavior: Piscivore/ Aquatic

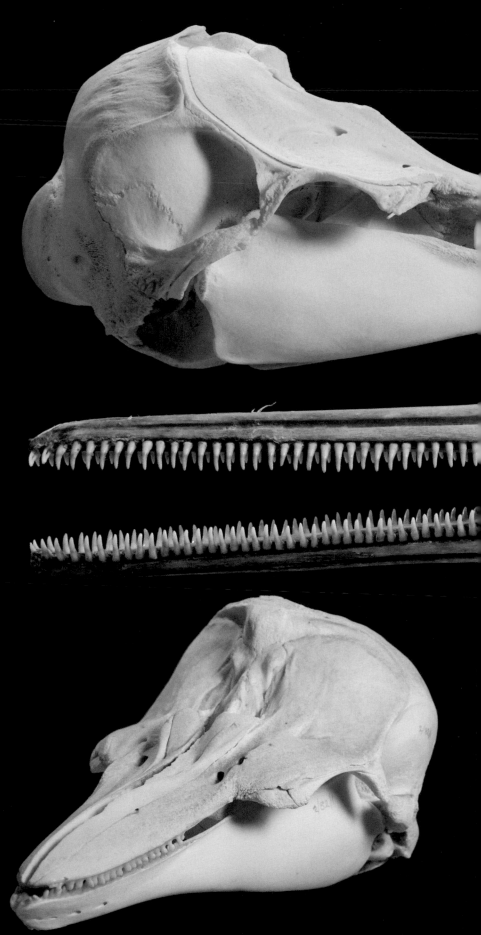

La Plata Dolphin
▷ *Pontoporia blainvillei*

THERE ARE, OR WERE, four species of river dolphin until about 2006, when a Chinese dam on the Yangtze River wiped out the closely related baiji. This specimen hails from the La Plata River in South America—another place where dolphins are vulnerable. The skull is notable for its huge rostrum or beak with rows of almost identical teeth. The dolphin, like reptiles, is unable to chew its food and must swallow its prey whole. It is one of the smallest cetaceans, growing no longer than the height of a small human.

AKA: Franciscana
Kingdom: Animalia
Phylum: Chordata
Class: Mammalia
Order: Cetacea

Family: Iniidae
Genus: Pontoporia
Behavior: Piscivore/ Aquatic

Harbor Porpoise
▷ *Phocoena phocoena*

THIS SMALL CETACEAN'S skull is surprisingly concave when you compare it to the living animal. The hollow forms a platform for the "melon"—an oily organ that serves to amplify sound sent out when the porpoise is echolocating. There are no discernible orbits since sight is poor in whales and dolphins. The teeth are unspecialized and the rostrum (or beak) is relatively short for a cetacean.

Kingdom: Animalia
Phylum: Chordata
Class: Mammalia
Order: Cetacea

Family: Phocoenidae
Genus: Phocoena
Behavior: Piscivore/ Aquatic

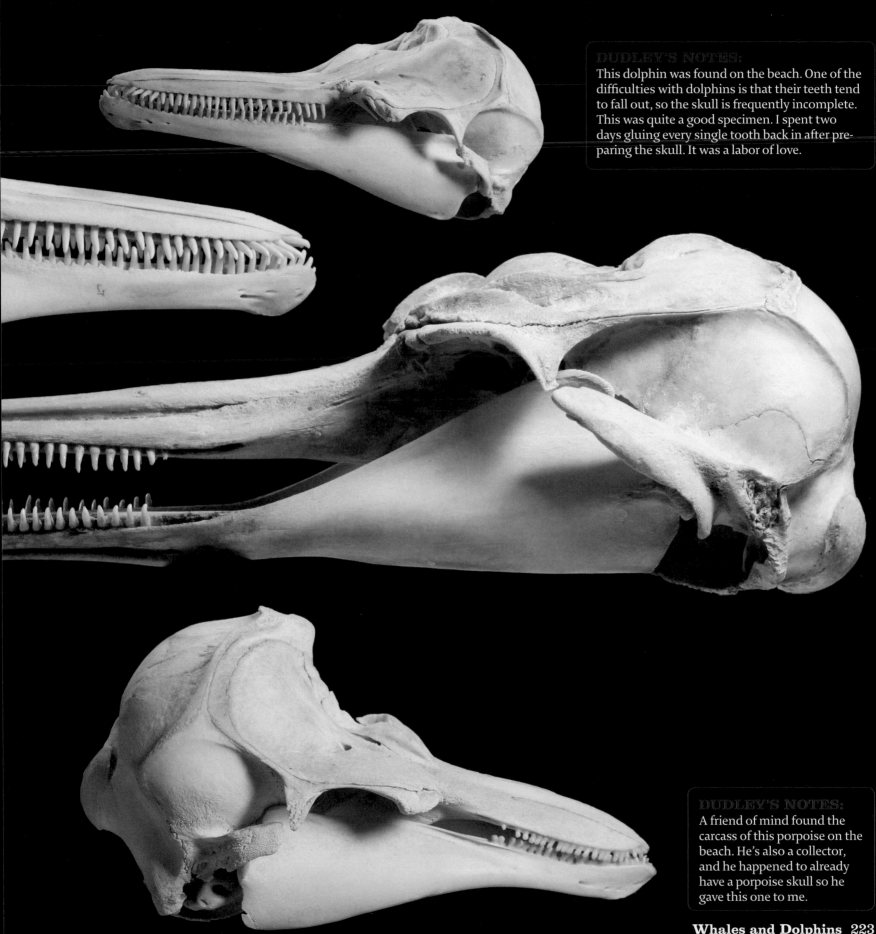

This dolphin was found on the beach. One of the difficulties with dolphins is that their teeth tend to fall out, so the skull is frequently incomplete. This was quite a good specimen. I spent two days gluing every single tooth back in after preparing the skull. It was a labor of love.

DUDLEY'S NOTES:

A friend of mind found the carcass of this porpoise on the beach. He's also a collector, and he happened to already have a porpoise skull so he gave this one to me.

Encounter with an Infamous Skull

IN JULY 1683 the Ottoman army from Istanbul famously laid siege to Vienna, the Turks under the command of a greedy, violent, drunken, sex-mad, and wildly xenophobic Turk with a fire-damaged face—the sultan's Grand Vizier, Kara Mustafa Pasha.

The siege did not quite work out in the Ottomans' favor. Quite to the contrary. The event is of historic importance because it represents the farthest that the Turks were ever to reach in their proposed expansion across Europe (they were heading for Rome, planning to consolidate their takeover of the Holy Roman Empire). The siege failed because of one almighty battle, in September, when there came entirely by surprise a massive flanking attack by cavalrymen from Emperor Leopold's ally, Poland. Vienna's misery was suddenly ended, and the Ottoman troops—300,000 of them, including some 20,000 elite Janissaries—were put to flight.

Kara Mustafa Pasha managed to halt his forces' retreat at Budapest, where he was met by appalled emissaries from the sultan's court who demanded (in the classical style of Turkish kismet) his immediate execution. The form of their message, though perhaps distorted down the ages, tells what happened next:

"Whereas for the Defeat of our Armies at the City of Vienna Thou Deservest to Die, it is Our Pleasure that Thou Entrust Thy Soul to the Ever-Merciful Lord, and that Thou Allow[est] to be Delivered Thy Head to these Our Messengers."

Kara Mustafa was then subject to the skilled attentions of the Ottoman court strangler, who used a bowstring. It took three uncomfortable minutes to kill the old man and then sever his head from his body. The head was skinned, stuffed, placed in a velvet bag, and sent down to Istanbul for Sultan Mehmet IV to inspect, then back to Budapest to be buried next to his body.

There the two disconnected entities might have stayed had not the Austrians turned up triumphant in Budapest five years later. The head—now no more than a hairless and fleshless skull—was exhumed, placed in a vitrine, brought back to Vienna, and put on display in the city museum. It has been there ever since.

It has been there, yes—but, since 1972, not on display. Modern Turkey complained to modern Austria about the human indignity of having the skull of one of their former leaders visible to the general public, and demanded it be returned to Turkey, to be reunited with the body (which had been brought to Istanbul three centuries before). The Austrian authorities, however, refused to send the skull back. They did agree to suspend its public display. So it was taken from its stand and placed in a warehouse some miles away, and was not seen publicly again.

Except that in 1999, I won permission to see it. It took a lot of doing: the museum director at first felt he must adhere to the rules that prevented any further tension between the former combatant powers of Turkey and Austria, now that the two countries were enwreathed in expressions of enthusiastic mutual amity.

But when I told him of my view—that the skull of Kara Mustafa symbolizes the ancient hostilities between the Ottoman and Austro-Hungarian empires—he agreed. He summoned an assistant who, he said, had brought the skull back from its warehouse of exile, just in case. When she arrived, he said softly to her: "Would you be so good as to take this gentleman"—gesturing at me—"to meet"—and here he pointed downstairs —"our other gentleman?"

I was taken to a basement room filled with the implements—bottles, knives, paint pots, brushes—of the picture restorers who normally live there. Sitting on a table that had been cleared for the occasion was a large cardboard box with lettering in ballpoint: Herr K. Mustafa. Hardly, I thought, in line with the expected dignity of a grand vizier of the Ottoman court.

An attendant sliced through the sealing tape, and lifted out the vitrine that held one of the most infamous skulls in the Balkans' long and skull-filled history.

The director's assistant gasped with muted delight. "I see him again!" she cried. "It has been so many years. I used to think of him almost as a friend."

He had probably never been a handsome man. His skull was brown and mottled. The eyes and nose sockets were large and deep, the orbits compressed into what must have been a permanent frown. There were five long teeth in the upper jaw, each yellowed and rotten

Portrait of the Grand Vizier Kara Mustafa Pasha, unknown artist, 1683. Oil on canvas.

and widely spaced. The entire lower jaw, the mandible, was missing. But there was also a length of finely made burgundy-colored cord. Had it been wrapped around the post on which the skull had been mounted? It had a tassel on one end. Or might this have been the bowstring with which the court strangler had choked the life out of Herr Mustafa?

I gazed at the skull for the next ten minutes or so. Outside it had been raining, but, as I held the vitrine up, the storm suddenly passed. A shaft of sunlight flooded into the room, lighting up the glass dome and its macabre inhabitant. It seemed a sign, a reminder. I handed the object back to its keeper, who laid it with some reverence back in its nest of tissue paper, closed the box lid over it, and sealed it away. Maybe it'll be seen in another thirty years or so. Maybe never.

A few days later I heard of the doings of a small academic industry devoted to studying the life and times of Kara Mustafa Pasha. It turned out that his headless body is now buried in northern Turkey, close to the Black Sea coastal town called Giresun, which in Roman times was Cerasus, where the first cherries are said to have come from. Symposia devoted to the grand vizier's life have been held there, and many of the Turkish attendees have expressed the hope that one day the skull in the Vienna vitrine can be brought back to the Black Sea coast and reburied to make the old man whole again.

There's a melancholy postscript to this story. We had the idea of photographing the grand vizier's skull, and asked permission from the museum authorities in Vienna. We had this reply—the language slightly polished—from Herr Direktor:

"I am sorry to tell you that we do not have the supposed skull of Kara Mustafa any more. It has been scientifically extremely disputed that this skull was indeed his skull—and so for this reason, and at least also because it is a matter of human dignity to bury the human remains of any unknown person, the skull was handed over to the urban cemetery board to be buried at the Wiener Zentralfriedhof. This happened in 2006."

Reptiles

American Alligator
Alligator mississippiensis ▷

THE AMERICAN ALLIGATOR is in many respects quite unlike the similarly sized Nile and saltwater crocodiles. A side-by-side comparison shows the alligator's snout is more rounded and its tooth line more even than the crocodile's. Alligators are also less aggressive by nature and tend to avoid human contact wherever possible. The photo shows a large adult in the process of eating a longnosed gar, the skull of which can be found elsewhere in the collection.

Kingdom: Animalia
Phylum: Chordata
Class: Reptilia
Order: Crocodylia

Family: Crocodylidae
Genus: Alligator
Behavior: Carnivore/
Aquatic

Dwarf Caiman
Paleosuchus palpebrosus ▽

THE DWARF CAIMAN is notable for its high forehead and large orbits when compared to other crocodilians. It is native to South America and is one of the smallest of the family.

AKA: Cuvier's Dwarf
Caiman, Musky Caiman
Kingdom: Animalia
Phylum: Chordata
Class: Reptilia
Order: Crocodylia
Family: Crocodylidae
Genus: Paleosuchus
Behavior: Carnivore/
Nocturnal

Spectacled Caiman
Caiman crocodilus ▷

THIS SKULL IS BELIEVED to have come from a spectacled caiman, a crocodilian from Central and South America. Its skull has the typical high forehead and large orbits of a caiman. Unlike many crocodile species, this one is relatively abundant, with an estimated one million in the wild.

AKA: White Caiman
Kingdom: Animalia
Phylum: Chordata
Class: Reptilia
Order: Crocodylia
Family: Crocodylidae
Genus: Caiman
Behavior: Carnivore/
Nocturnal

Nile Crocodile
Crocodylus niloticus ▷

THE NILE CROCODILE skull is one of the most menacing and impressive in the collection. All the teeth nestle together, interlocking to produce that irresistible and deathly grip. Only the protruding "canines" (a misnomer since only mammals possess such specialized teeth) break the uniform pattern. It is these, along with a narrower profile to the jaws, that help us distinguish, at a glance, crocodile skulls from those of alligators.

AKA: Common Crocodile
Kingdom: Animalia
Phylum: Chordata
Class: Reptilia
Order: Crocodylia

Family: Crocodylidae
Genus: Crocodylus
Behavior: Carnivore/
Nocturnal

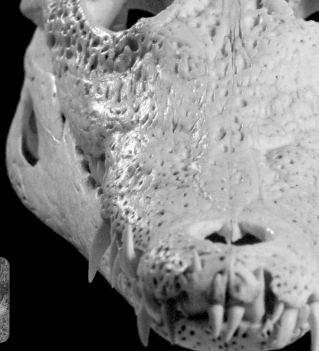

Saltwater Crocodile
Crocodylus porosus

THIS SALTWATER CROCODILE is the world's largest living reptile. This impressive skull is presented agape so that the teeth can be examined in all their glory. Crocodiles have a famously strong bite, but the muscles that open the mouth are weak, allowing their mouths to be taped shut for study or transport.

AKA: Estuarine Crocodile
Kingdom: Animalia
Phylum: Chordata
Class: Reptilia

Order: Crocodylia
Family: Crocodylidae
Genus: Crocodylus
Behavior: Carnivore/ Nocturnal

Central Bearded Dragon
Pogona vitticeps ▷

THESE LIZARDS hail from Western Australia and are popular with the pet trade. The skull is wedge-shaped with large orbits. The central bearded dragon can inflate its guttural pouch when threatened.

AKA: Inland
Bearded Dragon
Kingdom: Animalia
Phylum: Chordata
Class: Reptilia
Order: Squamata
Family: Agamidae
Genus: Pogona
Behavior: Omnivore/
Diurnal

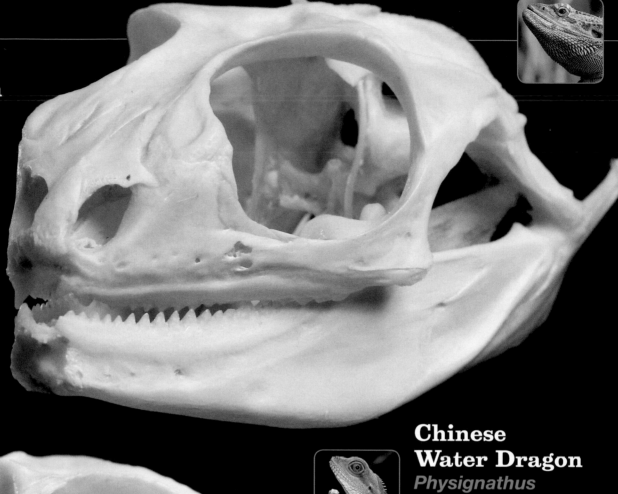

Chinese Water Dragon
Physignathus cocincinus ◁

WITH SUCH AN EVOCATIVE name, you might expect something special from the skull of a Chinese water dragon. It is notable among lizards for its large orbits (although Dudley has not preserved the sclerotic rings that would have made this specimen more recognizable). In the wild, when threatened, these dragons are known to drop into water and swim away at great speed.

AKA: Green Water Dragon
Kingdom: Animalia
Phylum: Chordata
Class: Reptilia
Order: Squamata
Family: Agamidae
Genus:Physignathus
Behavior: Insectivore/Diurna

Knight Anole
Anolis equestris ◁

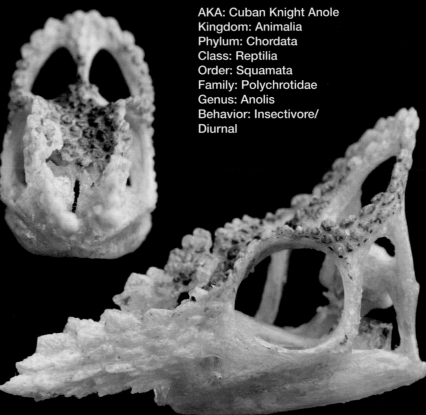

ANOLES HAVE DISTINCTIVELY pointed skulls. The knight anole (the biggest of the genus) has been endemic to Cuba, although there is a growing introduced population now thriving in Florida. In the wild this reptile is famously aggressive; it puffs itself up with air, gapes menacingly, and displays a skin flap (known as a dewlap) below its chin. This specimen, however, came from the pet trade, where they are quite docile and popular among collectors.

AKA: Cuban Knight Anole
Kingdom: Animalia
Phylum: Chordata
Class: Reptilia
Order: Squamata
Family: Polychrotidae
Genus: Anolis
Behavior: Insectivore/
Diurnal

Fischer's Chameleon
Bradypodion fischeri ▷

DUDLEY HAS A FEW CHAMELEONS in his collection, and they are very tricky to identify. This skull is of a Fischer's chameleon from East Africa. These are common in the pet trade, and there are a host of closely related species. The double-pronged rostrum on this small specimen is typical of the family. It is a lovely, almost jewel-like skull.

Kingdom: Animalia
Phylum: Chordata
Class: Reptilia
Order: Squamata
Family: Chamaeleonidae
Genus: Bradypodion
Behavior:
Insectivore/
Diurnal

Panther Chameleon
Furcifer pardalis ◁

THE PANTHER CHAMELEON is endemic to Madagascar, where this family of lizards is at its most diverse. Its casque is smaller than the impressive one sported by the veiled chameleon and is often pressed against the body, making it difficult to see. Panther chameleons have a flared bony trim to the upper jaw that extends from the eyes.

Kingdom: Animalia
Phylum: Chordata
Class: Reptilia
Order: Squamata
Family: Chamaeleonidae

Genus: Furcifer
Behavior: Insectivore/
Diurnal

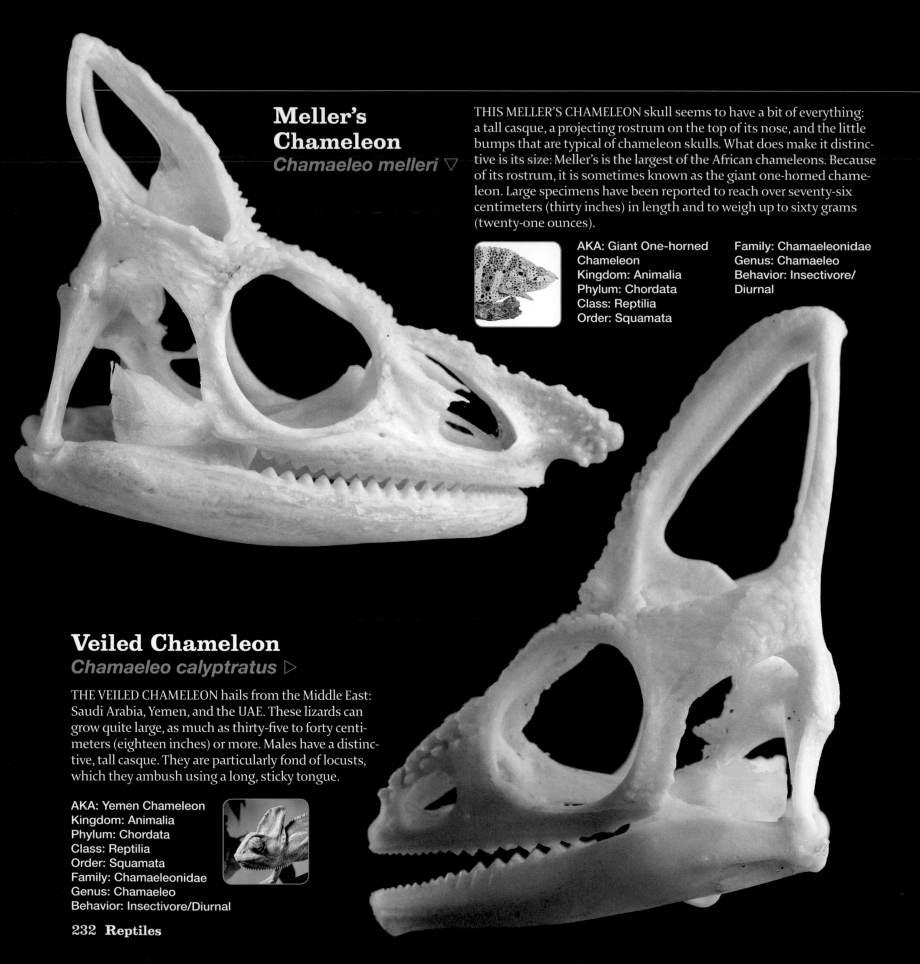

Meller's Chameleon
Chamaeleo melleri ▽

THIS MELLER'S CHAMELEON skull seems to have a bit of everything: a tall casque, a projecting rostrum on the top of its nose, and the little bumps that are typical of chameleon skulls. What does make it distinctive is its size: Meller's is the largest of the African chameleons. Because of its rostrum, it is sometimes known as the giant one-horned chameleon. Large specimens have been reported to reach over seventy-six centimeters (thirty inches) in length and to weigh up to sixty grams (twenty-one ounces).

AKA: Giant One-horned Chameleon
Kingdom: Animalia
Phylum: Chordata
Class: Reptilia
Order: Squamata

Family: Chamaeleonidae
Genus: Chamaeleo
Behavior: Insectivore/ Diurnal

Veiled Chameleon
Chamaeleo calyptratus ▷

THE VEILED CHAMELEON hails from the Middle East: Saudi Arabia, Yemen, and the UAE. These lizards can grow quite large, as much as thirty-five to forty centimeters (eighteen inches) or more. Males have a distinctive, tall casque. They are particularly fond of locusts, which they ambush using a long, sticky tongue.

AKA: Yemen Chameleon
Kingdom: Animalia
Phylum: Chordata
Class: Reptilia
Order: Squamata
Family: Chamaeleonidae
Genus: Chamaeleo
Behavior: Insectivore/Diurnal

Common House Gecko
Hemidactylus frenatus ▷

THE SKULL OF THE common house gecko is spare and simple. Its skull is flattened dorsoventrally, which helps it to keep a low profile when gripping smooth surfaces with its specially adapted feet. The small teeth are perfect for gripping insects that are attracted to the lights of houses—hence the name. You might see this familiar reptile scaling a wall or clinging onto a ceiling in warm countries.

Kingdom: Animalia
Phylum: Chordata
Class: Reptilia
Order: Squamata

Family: Gekkonidae
Genus: Hemidactylus
Behavior: Insectivore/
Nocturnal

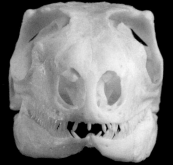

Green Iguana
Iguana iguana ▷

LIZARD TEETH VARY GREATLY depending on diet. The green iguana's are adapted for an herbivorous diet. Green iguanas can grow quite large for a lizard, and they are popular with the pet trade. Unusually among lizards (though quite commonly among iguanas), the green iguana is semiaquatic and is an accomplished swimmer.

AKA: Common Iguana
Kingdom: Animalia
Phylum: Chordata
Class: Reptilia
Order: Squamata

Family: Iguanidae
Genus: Iguana
Behavior: Herbivore/
Diurnal

Black Spiny-tailed Iguana
Ctenosaura similis

THE BLACK SPINY-TAILED iguana is found across Central and South America. This skull is not particularly distinctive (although its tiny teeth are numerous and sharp). When alive the black spiny-tailed iguana has the distinction of being the fastest living reptile, capable of speeds greater than twenty miles per hour. Juveniles are insectivorous, but adults prefer to eat leaves.

AKA: Black Iguana
Kingdom: Animalia
Phylum: Chordata
Class: Reptilia
Order: Squamata
Family: Iguanidae
Genus: Ctenosaura
Behavior: Omnivore/Diurnal

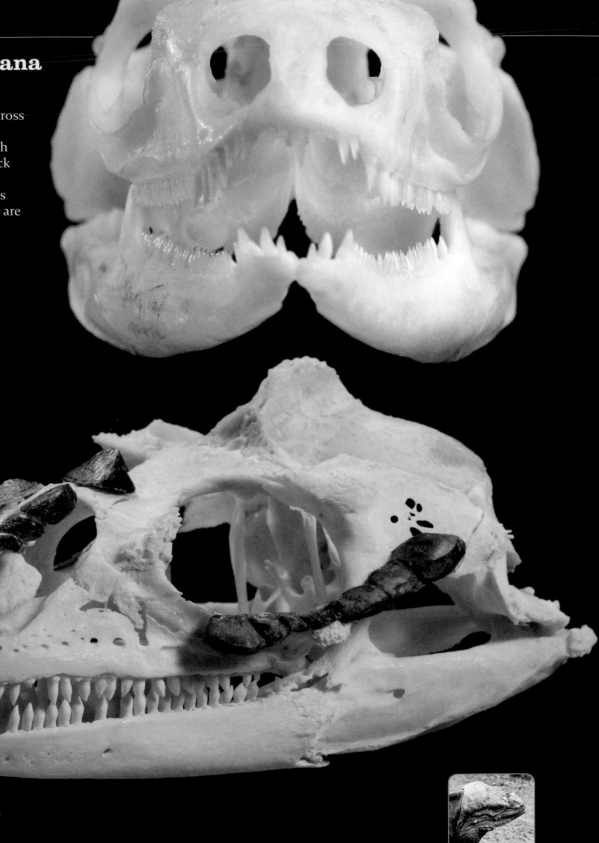

Rhinoceros Iguana
Cyclura cornuta

THIS IS A MOST impressive skull, notable for the keratinous growth on its zygomatic arches and the prominent tubercle developing between its eyes. The rhinoceros iguana is a native of the Caribbean. Males of the species fight each other, and the skull tubercles may act as defensive armor (this feature may also have a role in sexual selection). This skull doesn't show the large adipose pad that males develop on top of their heads, which may provide protection when fighting.

Kingdom: Animalia
Phylum: Chordata
Class: Reptilia
Order: Squamata

Family: Iguanidae
Genus: Cyclura
Behavior: Herbivore/Diurnal

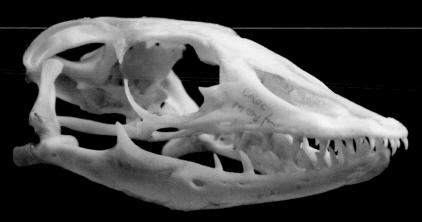

Emerald Tree Monitor
Varanus prasinus ◁

YOU'D BE FORGIVEN for mistaking this skull for that of a fish, perhaps a barracuda or pike. It is in fact a monitor, endemic to southern New Guinea, where it lives in trees and has a prehensile tail to help it climb. It is one of several tree-dwelling monitor species found on the Island of New Guinea.

AKA: Green Tree Monitor
Kingdom: Animalia
Phylum: Chordata
Class: Reptilia

Order: Squamata
Family: Varanidae
Genus: Varanus
Behavior: Carnivore/Diurnal

Water Monitor
Varanus salvator ▷

ALTHOUGH KNOWN AS the Malayan water monitor, this large lizard is in fact found throughout Southeast Asia. It has a rather menacing skull with well-spaced recurved teeth and a streamlined shape, and it is, as its name suggests, comfortable in the water. (It is easily confused with the crocodile monitor, to which it is closely related.)

AKA: Malayan Water
Monitor
Kingdom: Animalia
Phylum: Chordata
Class: Reptilia
Order: Squamata
Family: Varanidae
Genus: Varanus
Behavior: Carnivore/
Diurnal

Nile Monitor
Varanus niloticus ▷

THE STREAMLINED SKULL is a clue that this Nile monitor loves water. These lizards can be aged by their teeth: young animals have sharp pointed teeth; in older specimens the teeth are worn and rounded. Monitors can grow to a large size and are widespread across the African continent.

AKA: River Leguaan
Kingdom: Animalia
Phylum: Chordata
Class: Reptilia
Order: Squamata

Family: Varanidae
Genus: Varanus
Behavior: Carnivore/
Diurnal

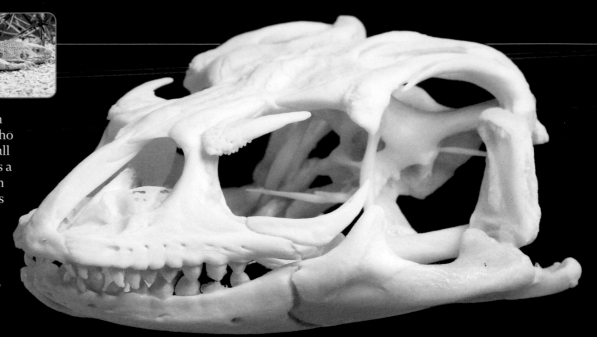

Savanna Monitor
Varanus exanthematicus

THE SAVANNA MONITOR is sometimes known as the Bosc's monitor after a French scientist who first described the species. It has a powerful skull with an impressive set of teeth. This animal has a cunning strategy to avoid being predated: when threatened, it seizes a rear leg in its mouth, rolls onto its back and plays dead; the resulting ring shape is almost impossible to swallow whole.

AKA: Bosc's Monitor
Kingdom: Animalia
Phylum: Chordata
Class: Reptilia
Order: Squamata

Family: Varanidae
Genus: Varanus
Behavior: Carnivore/
Diurnal

Burmese Python
Python molurus bivittatus

SNAKE SKULLS, particularly those of a python, are loosely connected multitudes of small bones that barely seem to touch each other and, unlike those of mammal skulls, are not joined by sutures. Pythons have small braincases, and the jaws dominate their skulls. Because their method of feeding includes swallowing large items of food whole (often items larger than the head itself), the Burmese python's upper and lower jaws can be disarticulated. Their teeth are uniform and point backwards to prevent prey from wriggling free.

Kingdom: Animalia
Phylum: Chordata
Class: Reptilia
Order: Squamata
Family: Boidae

Genus: Python
Behavior: Carnivore/
Diurnal

Blue-tongued Skink
Tiliqua sp. ▷

BLUE-TONGUED SKINKS come from Australasia. Their skulls are notable among lizards for their scaly appearance and are sought after among skull collectors. Their extraordinary blue tongues are thought to be a defense mechanism. When threatened, they gape widely and stick out their tongues.

Kingdom: Animalia
Phylum: Chordata
Class: Reptilia
Order: Squamata

Family: Scincidae
Genus: Tiliqua
Behavior: Omnivore/
Diurnal

Solomon Islands Skink
Corucia zebrata ◁

AS SKINK SKULLS GO, this is as big as they come: from nose to the tip of their tail, Solomon Island skinks can reach a length of eighty-one centimeters (thirty-two inches). This skink is a herbivore and doesn't have the sharply pointed teeth that many other lizards have. In their native Solomon Islands these lizards are under threat both due to habitat destruction and because they are hunted for food. This species is protected by CITES.

AKA: Prehensile-tailed
Skink
Kingdom: Animalia
Phylum: Chordata
Class: Reptilia

Order: Squamata
Family: Scincidae
Genus: Corucia
Behavior: Herbivore/
Nocturnal

Armadillo Lizard
Cordylus cataphractus

THIS IS HOW you might imagine a dragon's skull should look. The skull appears to be comprised of bony scales with remarkable rear-facing spines. In the wild, this southern African animal lives colonially in burrows. The armadillo lizard feeds on small invertebrates and, occasionally, small mammals. This is, unmistakably, a reptile skull—and a very attractive one at that!

AKA: Typical Girdled Lizard
Kingdom: Animalia
Phylum: Chordata
Class: Reptilia

Order: Squamata
Family: Cordylidae
Genus: Cordylus
Behavior: Insectivore/ Diurnal

Gila Monster
Heloderma suspectum

THE GILA MONSTER is large, squat, and slow-moving with a nasty and poisonous bite that has earned it a fearsome reputation. It seldom eats—less than once a month, usually—but when it does, it tends to devour its prey (lizards, frogs, small birds) headfirst and in one gulp. The skull of a gila monster is flattish and broad, covered with strange epidermal knobs; it has large nasal sockets and only slightly smaller eye sockets. Its teeth, all retrorse, are sharp and needle-like.

Kingdom: Animalia
Phylum: Chordata
Class: Reptilia
Order: Squamata
Family: Helodermatidae
Genus: Heloderma
Behavior: Carnivore/
Subterranean

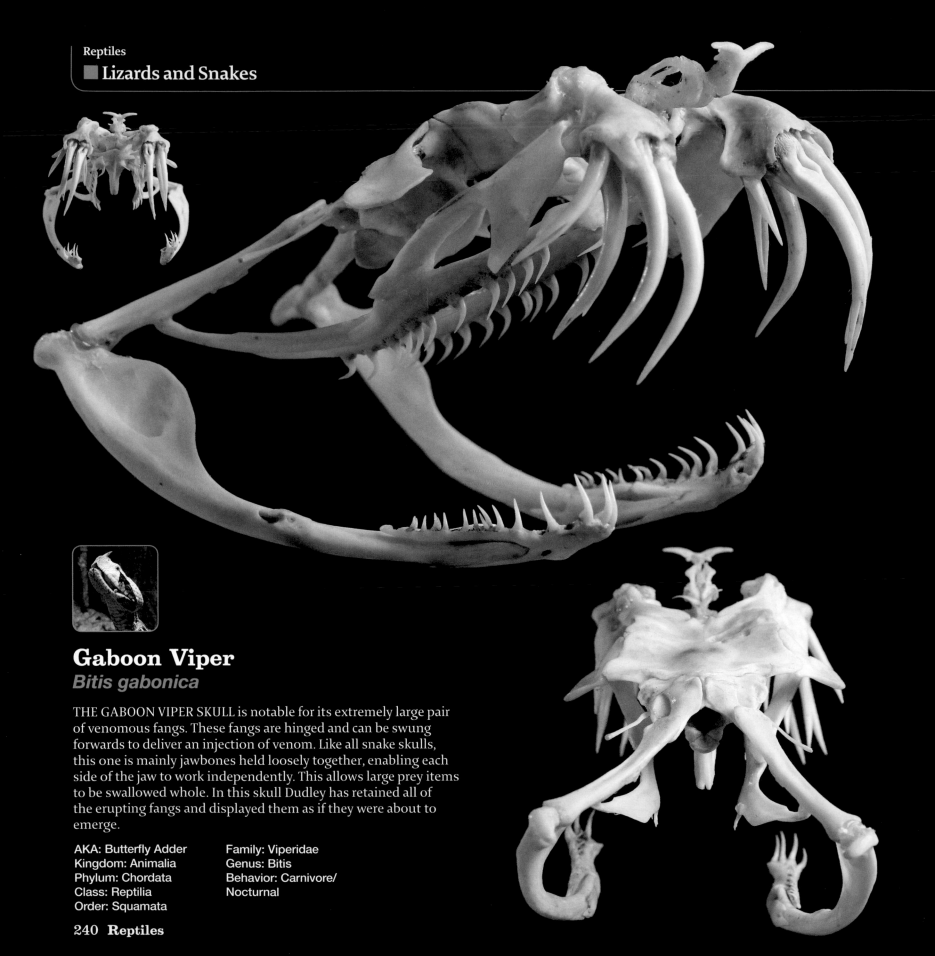

Gaboon Viper
Bitis gabonica

THE GABOON VIPER SKULL is notable for its extremely large pair of venomous fangs. These fangs are hinged and can be swung forwards to deliver an injection of venom. Like all snake skulls, this one is mainly jawbones held loosely together, enabling each side of the jaw to work independently. This allows large prey items to be swallowed whole. In this skull Dudley has retained all of the erupting fangs and displayed them as if they were about to emerge.

AKA: Butterfly Adder
Kingdom: Animalia
Phylum: Chordata
Class: Reptilia
Order: Squamata

Family: Viperidae
Genus: Bitis
Behavior: Carnivore/
Nocturnal

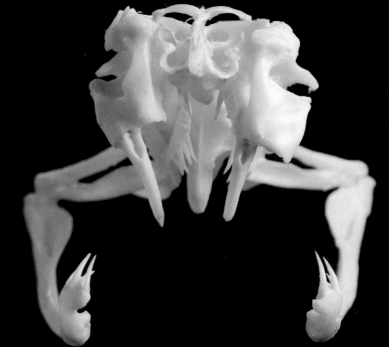

Western Diamondback Rattlesnake
Crotalus atrox

SNAKE SKULLS are truly fascinating things. In this rattlesnake skull, the fangs are displayed pointing backwards as they would be arranged before prey could be swallowed. Note the intricate bone structure in the nasal region: many snakes hunt by following scent trails.

Kingdom: Animalia
Phylum: Chordata
Class: Reptilia
Order: Squamata
Family: Viperidae
Genus: Crotalus
Behavior: Carnivore/
Nocturnal

Tegu
Tupinambis sp.

THE TEGUS OCCUPY a similar ecological niche in South America to that of the monitors in Africa and Southeast Asia, and their skulls are worth comparing if you're interested in convergent evolution. Tegus are carnivorous. Escaped populations now live in Florida—a place where all sorts of alien escapees from the pet trade seem to thrive.

Kingdom: Animalia
Phylum: Chordata
Class: Reptilia
Order: Squamata
Family: Teiidae
Genus: Tupinambis
Behavior: Omnivore/
Diurnal

Green Turtle
Chelonia mydas

THE GREEN TURTLE is the largest and one of the most widespread turtle species. Its skull is relatively featureless—there's no evidence of the keratinous sheath that lines the beak of some of the other turtle and tortoise jaws in the collection. The two cavities to the rear of the skull, bisected by the foramen magnum (where the spinal column would join), are typical of turtle skulls.

Kingdom: Animalia
Phylum: Chordata
Class: Reptilia
Order: Testudines
Family: Cheloniidae
Genus: Chelonia
Behavior: Herbivore/
Aquatic

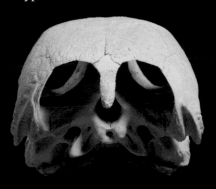

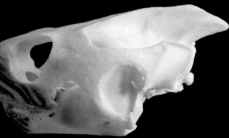

Chicken Turtle
Deirochelys reticularia

THE CHICKEN TURTLE seems to enjoy spending time on land, and it will wander quite a distance between ponds. This skull retains a beautiful tortoiseshell pattern on its keratinous beak.

Kingdom: Animalia
Phylum: Chordata
Class: Reptilia
Order: Testudines
Family: Emydidae
Genus: Deirochelys
Behavior: Omnivore/Diurnal

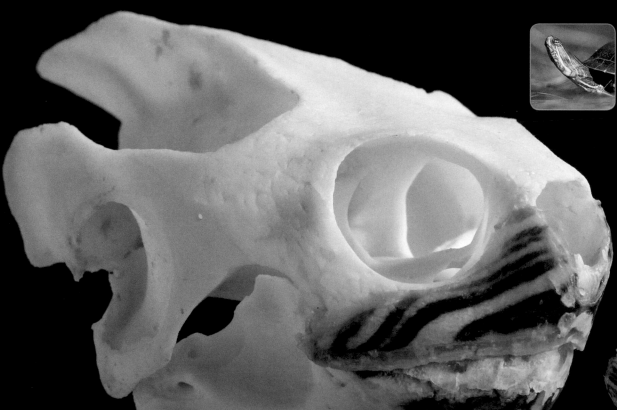

Mata Mata
Chelus fimbriatus ◁

THIS IS A MOST PECULIAR squat skull that bears the hallmarks of the turtle family: a beaklike jaw with a keratin sheath and the double opening to the rear of the skull. In the flesh the mata mata looks like a piece of floating bark and has a strange protuberance from the end of its nose that it uses like a snorkel. Its broad, flat mouth is an adaptation to what is known as suction feeding: by opening its jaw quickly, this turtle creates a wave of low pressure that sucks in a volume of water containing passing food (seahorses feed in the same way).

Kingdom: Animalia
Phylum: Chordata
Class: Reptilia
Order: Testudines
Family: Chelidae
Genus: Chelus
Behavior: Carnivore/Nocturnal

Florida Softshell Turtle
Apalone ferox ▽

THIS IS ONE OF the most peculiar of the turtle skulls. When viewed head-on, it looks as if it has just applied lipstick to its smiling mouth. This Florida softshell turtle has the most unbeaklike of turtle jaws, and the long fin at the back of the skull is reminiscent of the stream-lined helmet used by Olympic cyclists. In the wild they inhabit the southeastern U.S., not just Florida.

Kingdom: Animalia
Phylum: Chordata
Class: Reptilia
Order: Testudines

Family: Trionychidae
Genus: Apalone
Behavior: Carnivore/Aquatic

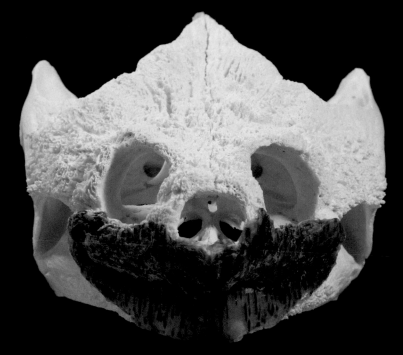

Common Snapping Turtle
Chelydra serpentina △

TURTLES HAVE VERY solidly constructed skulls. They have no teeth, but a horny beak (sometimes with serrations or ridges) covers their jaws, which are used for eating a wide variety of insects, shellfish, and other small aquatic animals.

Kingdom: Animalia
Phylum: Chordata
Class: Reptilia
Order: Testudines

Family: Chelydridae
Genus: Chelydra
Behavior: Carnivore/
Aquatic

Alligator Snapping Turtle
Macrochelys temminckii

THE ALLIGATOR SNAPPING turtle is named not for its jaws but for the alligator-like ridged plates on its back. These turtles have a fearsome reputation and can grow very large indeed—up to as much as 180 kilograms (400 pounds). Although long since lost to this skull, the tip of the turtle's tongue is shaped like a worm that is wiggled around enticingly to lure prey towards the beaklike jaws that, some claim, have the strongest biting force of any animal.

Kingdom: Animalia
Phylum: Chordata
Class: Reptilia
Order: Testudines
Family: Chelydridae
Genus: Macrochelys
Behavior: Piscivore/
Aquatic

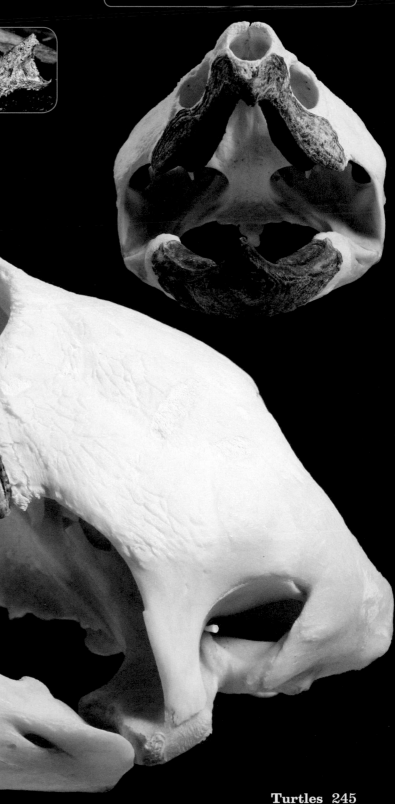

Aldabra Giant Tortoise

Dipsochelys dussumieri

IT'S NOT JUST THE GALAPAGOS Islands that are home to giant tortoises. In the Indian Ocean, giant tortoises were once found on Mauritius, the Seychelles, and, in particular, Aldabra Island. The tortoise skull is not dissimilar to those of the turtles in Dudley's collection, although the keratinous plates that line the beak also extend over the head of this skull. These animals can live for hundreds of years (and may even be the longest-lived vertebrates), so this large skull could be very old indeed.

Kingdom: Animalia
Phylum: Chordata
Class: Reptilia
Order: Testudines

Family: Tesudinidae
Genus: Dipsochelys
Behavior: Herbivore/
Diurnal

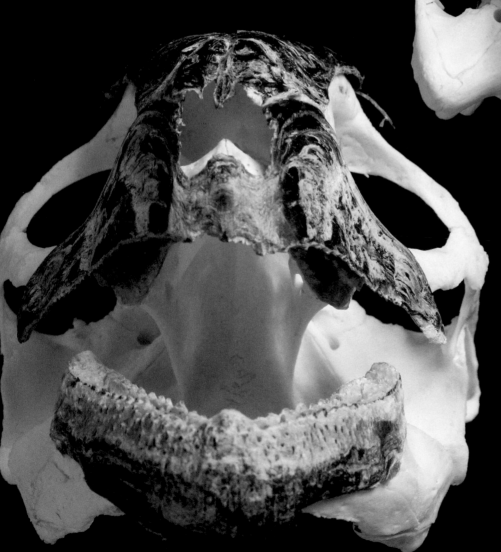

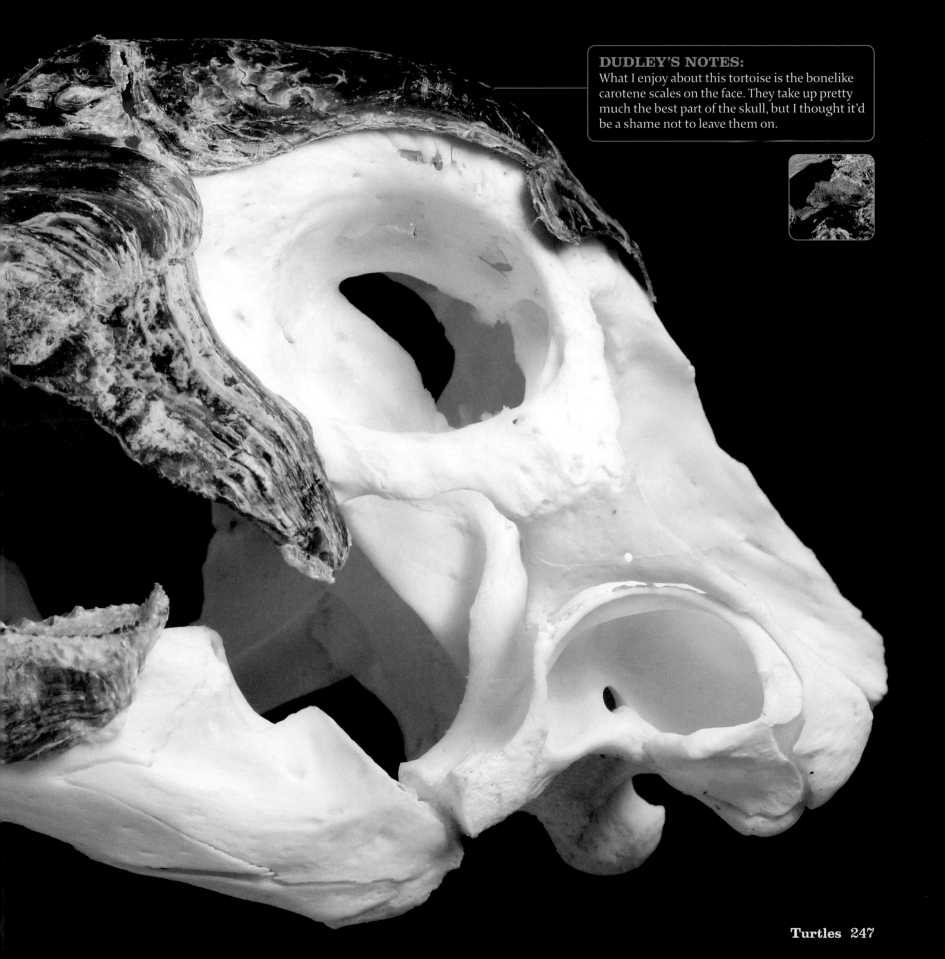

Indian Star Tortoise
Geochelone elegans

CLEARLY, THIS IS NOT A SKULL. In Dudley's collection room is a host of other animal bits and bobs: taxidermy, spider casts, hooves and teeth, and, in this case, a tortoise shell. The dorsal and ventral shields are known as a carapace (top) and a plastron (bottom). The Indian star tortoise's carapace has a series of bosses with radiating lines. Tortoises are the terrestrial members of an ancient reptile group that also includes turtles. They are very long-lived and, strangely, popular pets.

Kingdom: Animalia
Phylum: Chordata
Class: Reptilia
Order: Testudines

Genus: Geochelone
Behavior: Herbivore/ Diurnal

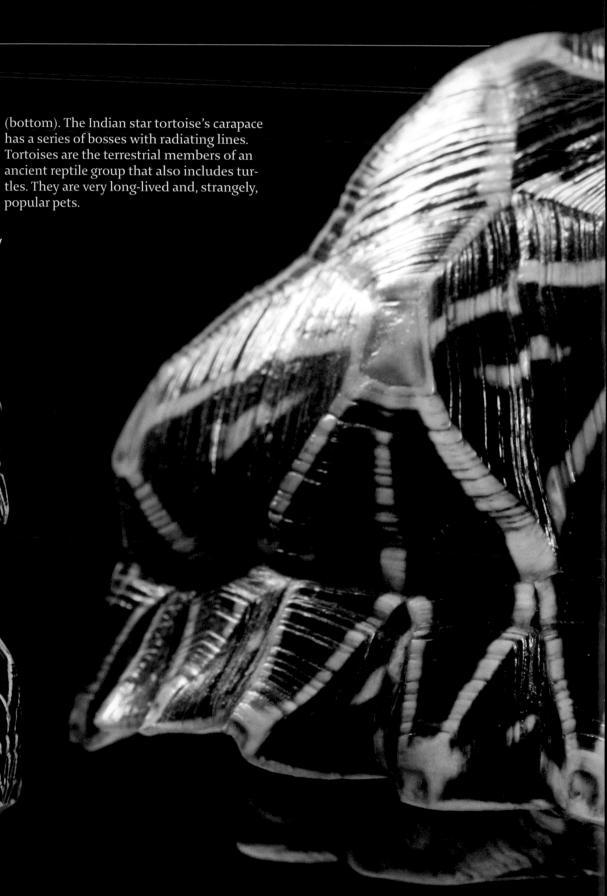

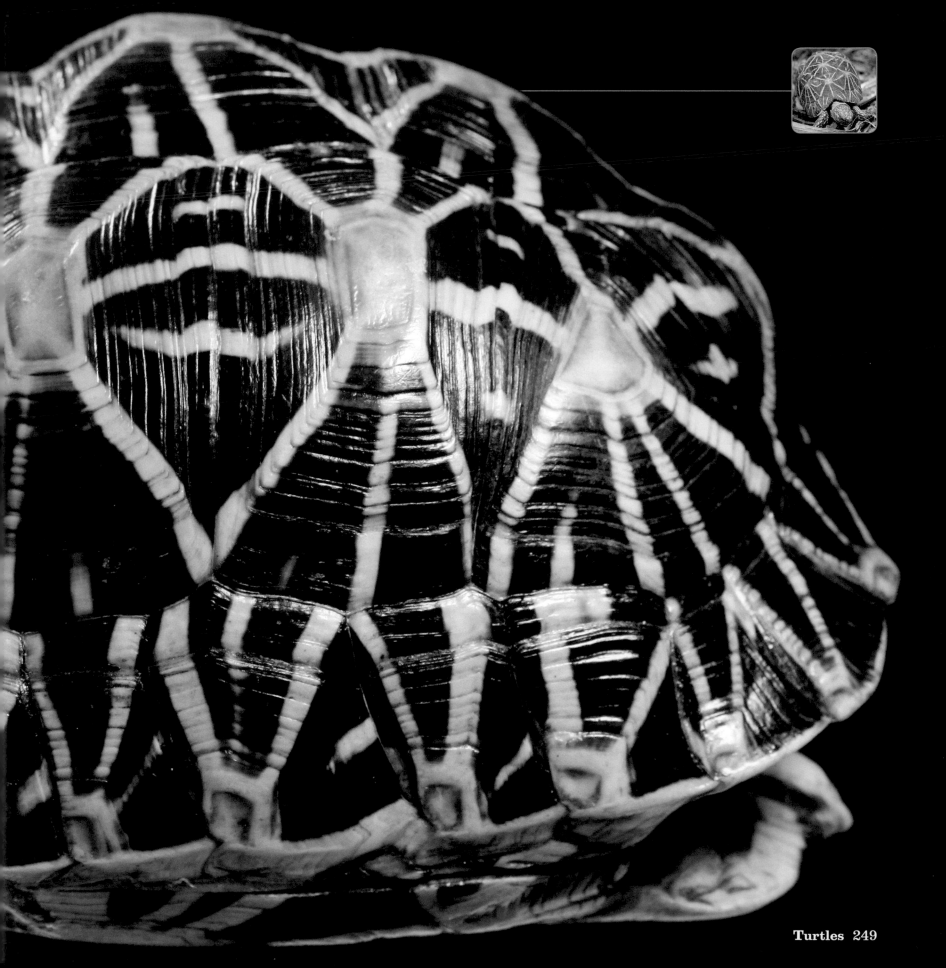

Photography Credits

istock photo: *Shofar,* 25

Naturepl.com
© Adriana Bacchello: *Rottweiler,* 155. © Alex Mustard: *California Sea Lion,* 158. © Andy Rouse: *Little Owl,* 73. © Andy Sands: *Fischer's Chameleon,* 231; *Common Shrew,* 195. © Ann & Steve Toon: *Springhare,* 215. © Anup Shah: *Giraffe,* 172; *Lesser Flamingo,* 87. © ARCO: *Domestic Pig,* 180; *Mute Swan,* 93. © Barry Mansell: *Alligator Snapping Turtle,* 245. © Bernard Castelein: *Hadada Ibis,* 89. © Brent Hedges: *Pygmy Marmoset,* 204. © Bruce Davidson: *Forest Buffalo,* 187. © Bruno D'Amicis: *Black Crowned Crane,* 87. © Christophe Courteau: *Hamadryas Baboon,* 205; *Hyacinth Macaw,* 76. © Claudio Contreras: *Common Vampire Bat,* 139. © Colin Varndell: *Gray Squirrel,* 216. © Daniel Heuclin: *Dwarf Caiman,* 226; *Helmeted Curassow,* 53. © Dave Bevan: *Domestic Sheep,* 187. © Dave Watts: *Northern Hairy-nosed Wombat,* 194; *Chilean Flamingo,* 87; *Duck-billed Platypus,* 171. © David Noton: *Vervet Monkey,* 209. © De Meester / ARCO: *Long-tailed Chinchilla,* 211; *Least Weasel,* 169. © Delpho / ARCO: *Common Raven,* 78. © Dietmar Nill: *Common Kestrel,* 48. © Dirscherl Reinhard: *Bowfin,* 119. © Doug Allan: *Harp Seal,* 160; *Hooded Seal,* 160. © Doug Perrine: *Mahi Mahi,* 110. © Doug Wechsler: *Brown Four-eyed Opossum,* 196. © Edwin Giesbers: *Clouded Leopard,* 147; *Leopard,* 147; *Lesser Short-nosed Fruit Bat,* 137; *Mouflon,* 188; *Tiger,* 150; *White Stork,* 92. © Eric Baccega: *Asiatic Black Bear,* 141. © Florian Graner: *Common Ling,* 117. © Francois Savigny: *Spectacled Caiman,* 227. © Gabriel Rojo: *South American Fur Seal,* 158. © George Sanker: *North American Beaver,* 212. © Hugh Maynard: *Hammer-headed Bat,* 138. © Igor Shpilenok: *Wolverine,* 179. © Inaki Relanzon: *Common Tenrec,* 218. © Jane Burton: *Axolotl,* 18. © Jeff Rotman: *Barracuda,* 108. © Jim Clare: *Guinea Pig,* 215; *Greater Galago,* 197. © Jose B. Ruiz: *Conger Eel,* 116. © Jouan & Rius: *Dromedary Camel,* 173; *Red Kangaroo,* 193; *Red-and-Green Macaw,* 76. © Jurgen Freund: *Long-spine Porcupinefish,* 113. © Kevin Schafer: *Black-footed Albatross,* 36; *Southern Cassowary,* 71; *Stump-tailed Macaque,* 208. © Kim Taylor: *Cape Fur Seal,* 158. © Laurent Geslin: *African Elephant,* 220; *Ring-necked Parakeet,* 77. © Luiz Claudio Marigo: *Black-necked Aracari,* 67; *Kinkajou,* 165; *Maguari Stork,* 90; *Six-banded Armadillo,* 134. © Lynn M. Stone: *American Alligator,* 226. © Mark Bowler: *Red-bellied Piranha,* 113. © Mark Cawardine: *Common Snapping Turtle,* 244. © Mark Taylor: *White-nosed Coatimundi,* 164. © Martin Gabriel: *Saltwater Crocodile,* 229. © Matthew Maran: *American Black Bear,* 141. © Michael Hutchinson: *Capybara,* 214. © Michael Pitts: *Tree Pangolin,* 136. © Mike Wilkes: *White-necked Raven,* 79. © Niall Benvie: *White-tailed Eagle,* 46. © Nick Garbutt: *Hog Badger,* 167; *Red Ruffed Lemur,* 203. © Nick Gordon: *Mata Mata,* 244. © Patricio Robles Gil: *Himalayan Monal Pheasant,* 49. © Paul Hobson: *Muntjac,* 174; *Little Penguin,* 83; *Plate-billed Mountain Toucan,* 66. © Peter Cairns / 2020Vision: *Razorbill,* 39. © Phil Chapman: *Common Spotted Cuscus,* 193. © Philippe Clement: *Carrion Crow,* 78. © Reinhard / ARCO: *Brown Rat,* 213. © Ric Fontijn: *Spot-billed Toucanet,* 68. © Richard du Toit: *Black Wildebeest,* 184. © Rod Williams: *Aardvark,* 134; *Aardwolf,* 156; *Abyssinian Ground Hornbill,* 56; *South American Tapir,* 191; *Chinese Water Deer,* 175; *Patagonian Mara,* 215; *Rufous Hornbill,* 62; *Common Treeshrew,* 218. © Rolf Nussbaumer: *Raccoon,* 164. © Shattil & Rozinski: *American Badger,* 167. © Simon King: *Meerkat,* 161. © Simon Williams: *Asian Openbill,* 90. © Solvin Zanki: *Harbour Porpoise,* 222; *Bicoloured Leaf-nosed Bat,* 139. © Staffan Widstrand: *American Rhea,* 72; *Razor-billed Curassow,* 52. © Stephen Dalton: *Central Bearded Dragon,* 230; *Field Vole,* 214; *Senegal Galago,* 197. © Steven David Miller: *Long-nosed Potoroo,* 193. © Suzi Eszterhas: *Eastern Yellow-billed Hornbill,* 57. © Tim Laman: *Philippine Tarsier,* 198; *Rhinoceros Hornbill,* 61; *Tarictic Hornbill,* 62; *Wrinkled Hornbill,* 65. © Todd Pusser: *Chicken Turtle,* 243; *La Plata Dolphin,* 222. © Tom Vezo: *Fisher,* 169; *Gray Wolf,* 154. © Tom Walmsley: *Smooth Newt,* 19. © Tony Heald: *Nile Monitor,* 235; *Trumpeter Hornbill,* 63. © Troels Jacobsen/Arcticphoto: *Subantarctic Fur Seal,* 159. © Visuals Unlimited: *Burmese Python,* 236; *Silvery-cheeked Hornbill,* 63; *Black-headed Spider Monkey,* 209. © Wegner / ARCO: *Bonobo,* 199; *Chihuahua,* 152; *European Hare,* 210. © Wild Wonders Of Europe/Lundgren: *Monkfish,* 118. © William Osborn: *Red-billed Hornbill,* 59

Steve Seal/BirdGuides
Steve Seal: *Great Cormorant,* 41

WikiCommons
663highland: *Raccoon Dog,* 154. A. Kniesel: *Ostrich,* 72. Alan D. Wilson: *Surf Scoter,* 94; *Polar Bear,* 140. Alex Dunkel: *Ring-tailed Lemur,* 203. Alexdi: *Hippopotamus,* 177. Althepal: *Double-toothed Barbet,* 69. AmericanXplorer13: *Green Iguana,* 233. Andreas Trepte: *Pied Avocet,* 92. Andrei Stroe: *Green Woodpecker,* 69. Art G.: *Mountain Lion,* 149. Badger Hero: *European Badger,* 167. Bertram Lobert: *Southern Brown Bandicoot,* 135. Bob Fabry: *Helmeted Guineafowl,* 49. Carl D. Howe: *North American Bullfrog,* 16. Catherine Trigg: *European Otter,* 168. Ceasol (Flickr): *American Bison,* 187. Chad Bordes: *Jabiru,* 90. Cody Pope: *Virginia Opossum,* 196. Colin M.L. Burnett: *African Bushpig,* 181. ©Larry D. Moore: *European Rabbit,* 210. D. Gordon and E. Robertson: *Warthog,* 181. Dûrzan Cîrano: *European Nightjar,* 70. Dacoman: *Great White Pelican,* 95. Dan Leo: *Gaboon Viper,* 240. Dan&Lin Dzurisin: *Striped Skunk,* 166. Derkarts: *Rock Hyrax,* 219. Doug Janson: *Helmeted Hornbill,* 59; *Lord Derby's Parakeet,* 77. Elektrofisch: *Budgerigar,* 75. Elf: *Boston Terrier,* 152. Eliezg: *Steller Sea Lion,* 159. Eric Kilby: *Andean Condor,* 47. fotokraj2: *Pesquet's Parrot,* 76.

Frank Wouters: *Red-handed Tamarin,* 204. GalawebDesign: *Pekingese Dog,* 155. Gibeah: *European Hedgehog,* 192. Hans Hillewaert: *Lowland Paca,* 213; *European Plaice,* 107. Henrik Thorburn: *Atlantic Puffin,* 39. Hugh Lunnon: *Marabou Stork,* 91. Ianaré Sévi: *Knight Anole,* 231. Ikiwaner: *Spotted Hyena,* 156. J.M. Garg: *Black Kite,* 45. John Hritz: *African Pied Hornbill,* 56. John Picken: *Great Northern Diver,* 93. Johnskate17: *Florida Softshell Turtle,* 244. Jojo: *Roe Deer,* 174. Jon Hanson: *Toco Toucan,* 68. Kalyanvarma: *Great Hornbill,* 58. Kamil: *European Eagle Owl,* 73. Keven Law: *Binturong,* 161; *European Red Fox,* 153; *Secretary Bird,* 48. Kyle Flood: *Alpaca,* 173. Liam Quinn: *Gentoo Penguin,* 83; *King Penguin,* 82. Lip Kee Yap: *Philippine Flying Lemur,* 192; *Spectacled Spiderhunter,* 81. Lipton Sale: *Blue Crowned Pigeon,* 84. Malene Thyssen: *Mandrill,* 206; *Orangutan,* 202. Marcel Burkhard: *Chinese Water Dragon,* 230. Markus Kolijonen: *Kea,* 74. Martin Sordilla: *Dwarf Cassowary,* 71. Masteraah: *North Sulawesi Babirusa,* 179. Melanie Milliken: *Argentine Horned Frog,* 17. Michael David Hill: *European Mole,* 195. Michael Ströck: *Giant Moray,* 116. Mila Zinkova: *Green Turtle,* 243; *Southern Giant Petrel,* 42. Mjobling: *White-chinned Petrel,* 42. Nester Galina: *Southern Sea Lion,* 157. Nick Hobgood: *Broadbarred Firefish,* 113; *Seahorse,* 106; *Trumpetfish,* 107. Noodle snacks: *Laughing Kookaburra,* 54. Nur Hussein: *Water Monitor,* 235. Patrick Coin: *Red-legged Seriema,* 49. Patrick Gijsbers: *Asian Small-clawed Otter,* 168. Pau Artigas: *Common Swift,* 70. Peter Massas: *Southern Ground Hornbill,* 64. Philippe Guillaume: *Gray Triggerfish,* 112. Pranav Yaddanapudi: *Blackbuck,* 185. Quartl: *Coypu,* 211. Richard Bartz: *Capercaillie,* 51. Rui Ornelas: *Burchell's Zebra,* 190. Ryan E. Poplin: *Black-and-white Colobus,* 208; *Sun Bear,* 142. Samuel Blanc: *Turkey Vulture,* 48. Sander van der Wel: *Wreathed Hornbill,* 64. Sarah McCans: *Nile Crocodile,* 227. Spencer Wright: *Common Buzzard,* 44. Stan Shebs: *Armored Catfish,* 114; *Southern Rockhopper Penguin,* 83. Stephen Hanafin: *Blyth's Hornbill,* 57. Steve Childs: *Warty Frogfish,* 117. Steve Garvie: *Lappet-faced Vulture,* 44; *Saddle-billed Stork,* 91; *Tufted Grey Langur,* 208; *White-bellied Go-away-bird,* 81. Steven G. Johnson: *Longnose Gar,* 114. Stevie B (Flickr): *Barn Owl,* 73. su neko: *Green Aracari,* 67; *Siamang,* 199. Sylfred: *Savanna Monitor,* 236. TigerhawkVox: *Western Diamondback Rattlesnake,* 241. Tim Parkinson: *Fennec Fox,* 153. Tim Vickers: *Domestic Cat,* 146. Tino Strauss: *Atlantic Pollock,* 117. Tom Freidel: *Tegu,* 242. Tomascastelazo: *La Caterina,* 130. Trisha Shears: *Potto,* 198. Viborg: *Great Dane,* 154. Wilfried Berns: *Armadillo Lizard,* 238. Winifried Bruenken: *Kori Bustard,* 52. yaaaay (Flickr): *African Lion,* 144.

Shoebill, 88. *Woodchuck,* 216. *Hamlet and Horatio in the Cemetery,* 1839, Eugène

Delacroix, 122. *De Humani corporis fabrica,* 1543, Andreas Veslaius, 122. Print engraving of Stede Bonnet in Charles Johnson's *A General History of the Pyrates,* c. 1725, 124. *The Coat of Arms with the Skull,* Albrecht Durer, 1507, 126. *Skull with a burning cigarette.* 1885 or early 1886. Oil on Canvas. Vincent van Gogh, 127. *Two Skulls in a Window Niche,* 1520. Oil on lime wood. Hans or Ambrosius Holbein, 127. *Portrait of the Grand Vizier Kara Mustafa Pasha,*1683. Oil on canvas, 224.

Andreas Trepte, www.photonatur.de: *Northern Shoveler,* 94; *Eurasian Spoonbill,* 88; *Great Black-backed Gull,* 41; *Harbor Seal,* 160

Aurélien AUDEVARD: *White-winged Grosbeak,* 80

Bart Hazes: *Redtail Parrotfish,* 110

Ben Twist: *Blue-tongued Skink,* 237

Bill Schmoker: *Chinese Goose,* 93

Dave Switzer: *Mountain Beaver,* 212

Dr. Heike Lutermann: *Cape Mole Rat,* 211

Eric Isselée: *Meller's Chameleon,* 232

Fergus Kennedy: *Houndfish,* 118

Joachim S. Müller: *Atlantic Wolffish,* 111

Johannes Pfleiderer: *Yellow-casqued Wattled Hornbill,* 65

Johnny Sandaire: *American Blackbelly,* 186

Ken Billington: *Great Crested Grebe,* 95

Piotr Jonczyk: *Brown-cheeked Hornbill,* 57

Robert Fenner: *Starry Triggerfish,* 112 Sergey Sosnovskiy: *Roman memento mori from Pompeii,* 122

Setsuko Winchester: *Human,* 200

Silent Kid (Flikr): *Cheetah,* 148

B. Peterson: *Kit Fox,* 153

Dr. Dwayne Meadows: *Black Jack,* 110

Jeff Servoss: *Gila Monster,* 239

NASA: *Common Bottlenosed Dolphin,* 222

© WWF-Greater Mekong: *Javan Rhinoceros,* 191

A Skull Sectioned, 1489. Leonardo da Vinci. Reproduced with gracious permission of Her Majesty The Queen. The Royal Library, The Royal Collection, Windsor Castle, 126

Anterior

The front or head end of an animal.

Antler

An appendage, often branching, grown annually on the top of the head of many deer species.

Auditory bullae

The hollow structures on either side of the skull enclosing the middle and inner ear.

Canines

Often long and pointed teeth, used to grip and tear at food, especially meat. Enlarged in some species, they can also form tusks, for example in the walrus.

Carapace

The dorsal section of the shell or exoskeleton.

Carnassials

Large teeth found in many carnivorous mammals, usually the last premolar and first molar, used for shearing flesh and bone.

Carnivore

An organism whose diet consists largely or solely of meat.

Casque

An enlargement on the upper mandible of some bird species, such as hornbills, or a large growth on the skull of cassowary species.

CITES

Convention on International Trade in Endangered Species of Wild Fauna and Flora, also known as the Washington Convention. A treaty drafted to ensure international trade in wild animal and plant species does not threaten the survival of a species in the wild.

Cranium

The main body of the skull; the entire structure less the lower mandible.

Dental formula

A standardized method for presenting the number and type of teeth present in the skull of any given mammal.

Dentine

A mineralized tissue found in mammalian teeth, coated by enamel and forming part of the root below.

Diastema

A gap between teeth.

Dorsoventral

The plane of reference running from the back to the belly of an animal, perpendicular to the anterior-posterior plane of reference that runs from the head to the back end of the animal.

Enamel

A highly mineralized tissue that coats the exposed surface of mammalian teeth.

Endemic

Limited in distribution to a defined geographical area.

Fontanel

Soft areas of young mammalian skulls, allowing the skull of an infant to flex during birth. The fontanel hardens to form sutures.

Foramen magnum

An opening at the base of the skull through which the brain stem passes, linking the brain to the spinal cord.

Frugivore

An organism whose diet consists largely or solely of fruit.

Herbivore

An organism whose diet consists largely or solely of plant matter.

Incisors

The front teeth, used for nipping and shearing. Enlarged in some species, they can also form tusks, for example in elephants and narwhals.

IUCN

International Union for Conservation of Nature. The conservation body and

Index